Thomas A. Willging

On Length Spectra of Lattices

On Length Spectra of Lattices

by
Thomas A. Willging

Dissertation, Karlsruher Institut für Technologie
Fakultät für Mathematik, 2010

Impressum

Karlsruher Institut für Technologie (KIT)
KIT Scientific Publishing
Straße am Forum 2
D-76131 Karlsruhe
www.ksp.kit.edu

KIT – Universität des Landes Baden-Württemberg und nationales
Forschungszentrum in der Helmholtz-Gemeinschaft

KIT Scientific Publishing 2010
Print on Demand

ISBN 978-3-86644-584-0

On Length Spectra of Lattices

Zur Erlangung des akademischen Grades eines

DOKTORS DER NATURWISSENSCHAFTEN

von der Fakultät für Mathematik des

KIT
(Karlsruher Institut für Technologie)

genehmigte

DISSERTATION

von

Thomas A. Willging
aus Stuttgart

Tag der mündlichen Prüfung:	16. November 2010
Referent:	PD Dr. Stefan Kühnlein
Korreferent:	Prof. Dr. Frank Herrlich

Preface

The theory of quadratic forms is a subject in number theory of the purest sort going back to Fermat, Euler, Lagrange, Gauß and Minkowski to mention but a few. By a quadratic form Q in dimension n we mean a function

$$Q(x) = \sum_{i,j=1}^{n} a_{ij} x_i x_j$$

of degree 2 with $x := (x_1, ..., x_n)^T \in \mathbb{R}^n$ and coefficients $a_{ij} = a_{ji} \in \mathbb{R}$.
An old natural question is to ask which integers are represented over $\mathbb{Z}$ by a given form with integer coefficients and moreover in how many ways they are represented. In some instances these questions can be answered but there is no general answer. The central idea is the so called local-global principle, that means we first check if an integer is represented locally over all $\mathbb{Z}_p$ and over $\mathbb{R}$.

As is well known, the theory of positive definite quadratic forms provides an alternative approach for studying lattices. There is a one-to-one correspondence between congruence classes of lattices and equivalence classes of quadratic forms. Therefore other classical questions as for instance the sphere packing problem or the kissing number problem can be considered in this context too.

The aim of this thesis is to study a related question, Schmutz Schaller's conjecture, that in dimensions 2 to 8 the quadratic forms associated with the lattices with the best known sphere packings are maximal. We say respectively that these lattices have maximal lengths. This means that their k-th length is strictly greater than the k-th length of any other lattice in the same dimension with the same covolume. Here it is important that we do not count the multiplicities of these lengths.

After we have introduced the basic concepts in Chapter 1 we will see in Chapter 2 that the Schmutz Schaller conjecture does not hold true for dimension 3. Although the statement holds asymptotically, i.e. if k is big enough, we will explicitly present a counter-example in Section 2.3. It turns out that only its 6-th length is not dominated by the corresponding length of the lattice with the best sphere packing (Proposition 2.3.5). However, it seems that there is nothing but this exception: one lattice, where for one length the conjecture fails. The results of Section 2.3 are published in [Wi].

To support the guess that there is only this counter-example we will discuss the conjecture for ternary lattices with bounded multiplicities. Here the main result is, that the conjecture applies for all these lattices with bounded multiplicities (Theorem 2.5.5).

It may be surprising that the multiplicities of the lengths do not play a role at all. This is contrary to the usual definition of the length spectrum. But in contrast to the situation mentioned before the lengths of the complete length spectrum have the same asymptotic behaviour for all lattices with fixed covolume.
We will use this fact to prove in Chapter 3 that a lattice with maximal complete lengths does not exist in any dimension $n \geq 2$. In particular we will prove that for any lattice there exists another lattice, arbitrarily close, such that its complete lengths are not dominated by the complete lengths of the first lattice (Theorem 3.2.4).

It is then natural to ask, whether there exist two lattices at all, such that the complete lengths of one lattice dominate the complete lengths of the other. It seems that this is a difficult question in general, but we will answer this question in the negative in the special case of even unimodular lattices in the last section.

At this place I would like to thank all persons who have contributed to this thesis, in particular:

PD Dr. Stefan Kühnlein for his guidance and helpful advice throughout the whole time, *Prof. Dr. Frank Herrlich* for his support and for kindly agreeing to be Korreferent of this thesis, *Ute Luhm* and *Lothar Lemp* for their careful proofreading.

Contents

Chapter 1

Basic concepts

1.1 Lattices and quadratic forms

A *lattice* Γ in the Euclidean standard space $\mathbb{R}^n$ is a subgroup which is generated by a basis $B = \{b_1, ..., b_n\}$.

$$\Gamma := \left\{ \sum_{i=1}^n x_i b_i : x_i \in \mathbb{Z} \right\}$$

The volume of the fundamental parallelotope $F_B := \{ \sum_{i=1}^n \alpha_i b_i : 0 \le \alpha_i < 1 \}$ of Γ is called the *covolume* of Γ.

As is well known, the theory of quadratic forms offers an alternative language for studying lattices. There is a one-to-one correspondence between congruence classes of lattices and equivalence classes of quadratic forms.

Definition 1.1.1.

(a) Let Γ be a lattice in $\mathbb{R}^n$ with basis $B = \{b_1, ..., b_n\}$ and define the matrix $M_B := (b_1|...|b_n)$. Then the quadratic form

$$Q_{\Gamma_B}(x) := x^T \cdot \underbrace{M_B^T M_B}_{=: A_{\Gamma_B} =: (a_{ij})} \cdot x,$$

$x \in \mathbb{Z}^n$, is *associated* with Γ. Obviously, the matrix A_{Γ_B} is symmetric and positive definite. Therefore by a "form" we always mean a positive definite quadratic form.

(b) Furthermore we call a form Q *integral* if $Q(x) \in \mathbb{Z}$ for all $x \in \mathbb{Z}^n$ and we call Q *classically integral* if $a_{ij} \in \mathbb{Z}$ for all $i, j \in \{1, ..., n\}$.

(c) Let I be a ring in a field. Two quadratic forms Q and Q' with matrices A and A' are called *(I-)equivalent* or in the same *class* if there exists a matrix $S \in I^{n \times n}$ with $\det(S) \in I^*$ such that $A = S^T A' S$.
For $I = \mathbb{Z}$ two forms are equivalent, if they refer to the same lattice with different bases.

To put a finer point on that, we exploit that the set of lattices in $\mathbb{R}^n$ is $\mathrm{GL}_n(\mathbb{Z})\backslash\mathrm{GL}_n(\mathbb{R})$ and the set of real quadratic forms is $\mathrm{GL}_n(\mathbb{R})/\mathrm{SO}_n(\mathbb{R})$. We can thus identify the set of congruence classes of lattices respectively of $\mathbb{Z}$-equivalence classes of quadratic forms with

$$\mathrm{GL}_n(\mathbb{Z})\backslash\mathrm{GL}_n(\mathbb{R})/\mathrm{SO}_n(\mathbb{R}).$$

According to this principle we call a lattice Γ *(classically) integral* if its associated form Q_Γ is (classically) integral, *even* if $Q_\Gamma(x) \in 2\mathbb{Z}$ for all $x \in \mathbb{Z}^n$ and *arithmetic* if Q_Γ is arithmetic, i.e. there exists a $\lambda \in \mathbb{R}$ such that $\lambda \cdot Q_\Gamma$ is integral. As well we define $\det(Q_\Gamma) := \det(A_\Gamma) = (\mathrm{cov}(\Gamma))^2$.

In the sequel we will switch freely between the language of lattices and the language of forms.

1.2   Reduction

By "reduction" we mean choosing a representative of a class with nice properties, i.e. whose coefficients satisfy certain inequalities, depending on the reduction. In this sense reduction for lattices consists in finding bases so that the scalar products of the elements satisfy these inequalities.

Definition 1.2.1. A positive definite form Q with matrix $A = (a_{ij})$ is said to be *reduced* (in the sense of Minkowski) if for $k = 1, ..., n$

$$a_{kk} \leq Q(x)$$

for all integral vectors $x := (x_1, ..., x_n)^T$ with $\gcd(x_k, ..., x_n) = 1$,
and if in addition $a_{1j} \geq 0$ for $j = 2, ..., n$.

In every class there exists a reduced form. More precisely we have, cf. [Ca, p.256]:

Theorem 1.2.2. *Every positive definite form is equivalent to at least one and at most finitely many reduced forms.*

It is clear that the reduction conditions imply that

$$a_{11} \leq a_{22} \leq ... \leq a_{nn}.$$

An additional consequence is

$$|2a_{ij}| \leq a_{ii}, \quad (1 \leq i < j \leq n).$$

We remark the fact that if $n \geq 3$ these consequences are not sufficient for a form to be reduced. However, these conditions will do all we need.

For an associated lattice with basis $\{b_1, ..., b_n\}$ it follows that

$$|\cos(\sphericalangle(b_i, b_j))| = \frac{|\langle b_i, b_j \rangle|}{|b_i| \cdot |b_j|} \leq \frac{\frac{1}{2}|b_i|^2}{|b_i| \cdot |b_i|} = \frac{1}{2} \quad (i \neq j).$$

Hence

$$-\frac{1}{2} \leq \cos(\sphericalangle(b_i, b_j)) \leq \frac{1}{2}, \text{ respectively } 0 \leq \cos(\sphericalangle(b_1, b_j)) \leq \frac{1}{2}.$$

1.3 Local considerations

An old problem in the theory of quadratic forms is the question of representing integers by a positive definite integral form. To that end we will use the so called *local-global principle*. That means we will first check if an integer m is *represented* by Q locally, i.e. whether $Q(x) = m$ with $x \in \mathbb{Z}_p^n$ is solvable at all places p, where p is either a prime number (for p-power congruences) or ∞ (for the sign, with the usual convention that $\mathbb{Z}_\infty = \mathbb{R}$). Unfortunately the very satisfactory local-global theory of forms over $\mathbb{Q}$ does not hold over $\mathbb{Z}$. While two forms are $\mathbb{Q}$-equivalent if and only if they are $\mathbb{Q}_p$-equivalent for all p (Weak Hasse Principle), there exist forms which are $\mathbb{Z}_p$-equivalent for all p but not $\mathbb{Z}$-equivalent, cf. examples in Section 2.4. Therefore we define:

Definition 1.3.1. The *genus* gen(Q) of an integral form Q is the set of all (integral) forms that are *locally equivalent* to Q, i.e. $\mathbb{Z}_p$-equivalent at all places p.

Obviously gen(Q) is a disjoint union of classes. Now we get the local-global connection:

Theorem 1.3.2. *Let Q be an integral form which represents an integer m locally. Then m is represented by one form in* gen(Q).

A proof can be found for instance in [Ca, Chap.9.5] or [Kn, Satz 22.1].
We make note of the trivial consequence:

Corollary 1.3.3. *If the genus of Q has only one class, then m is represented locally by Q if and only if m is represented by Q.*

Hence if an integral form Q has only one class in the genus, all numbers that are not represented by Q can be described by congruence conditions. From Definition 1.3.1 we see that for locally equivalent forms Q and Q' the rational number $\frac{\det(Q)}{\det(Q')}$ is a p-adic unit for all p, so it must be ± 1. Since all forms are assumed to be positive definite (or with $p = \infty$) it follows that $\det(Q) = \det(Q')$.
Furthermore it is well known that if a prime number p does not divide $2 \cdot \det(Q)$ there is only one $\mathbb{Z}_p$-equivalence class of forms of the same determinant (for simplicity we assume all forms to be classically integral):

Theorem 1.3.4. *Let Q, Q' be two classically integral forms such that $\det(Q) = \det(Q')$ is a p-adic unit for an odd prime p. Then Q and Q' are $\mathbb{Z}_p$-equivalent.*

Sketch of proof. One can show (cf. [Ca, p.116]) that Q is $\mathbb{Z}_p$-equivalent to a form

$$\sum_{i,j=1}^{n} b_{ij} x_i x_j,$$

where $|b_{ii}|_p \le |b_{11}|_p$ and (since $p \ne 2$) $|b_{ij}|_p \le |b_{11}|_p$. Therefore $\frac{b_{ij}}{b_{11}} \in \mathbb{Z}_p$ for all b_{ij} and we can complete the square. So Q is $\mathbb{Z}_p$-equivalent to a form

$$b_{11} x_1^2 + H(x_2, ..., x_n)$$

for some $(n-1)$-dimensional form H. By induction it follows that Q is $\mathbb{Z}_p$-equivalent to a diagonal form $b_{11}x_1^2 + \ldots + b_{nn}x_n^2$, where the b_{ii} are p-adic units since $p \nmid \det(Q)$. And such a form is integrally equivalent to

$$x_1^2 + x_2^2 + \ldots + x_{n-1}^2 + \det(Q)x_n^2.$$

$\square$

Hence there are only finitely many "critical" p to consider. At this place we will state (and also prove) a further consequence that will be helpful later on:

Proposition 1.3.5. *Let Q be an n-dimensional classically integral form, $n \geq 3$ and p an odd prime not dividing $\det(Q)$. Then Q represents all natural numbers m over $\mathbb{Z}_p$.*

This proposition can be verified with Hensel's Lemma in its simplest variant:

Lemma 1.3.6 (Hensel). *Let $f(x)$ be a polynomial in the single variable x and suppose that there exists an $x_0 \in \mathbb{Z}_p$ such that*

$$|f(x_0)|_p < |f'(x_0)|_p^2,$$

where $f'(x)$ denotes the (formal) derivative with respect to x. Then there is a $y \in \mathbb{Z}_p$ such that $f(y) = 0$.

For a proof one can refer to [Ca, p.47].

Corollary 1.3.7. *Let p be an odd prime and define the integral form $Q(x) := a_1x_1^2 + a_2x_2^2$ such that $a_1, a_2 \not\equiv 0 \, (\mathrm{mod} \, p)$. Then for all positive integers $b \not\equiv 0 \, (\mathrm{mod} \, p)$ there exists a $y \in \mathbb{Z}_p^2$ such that $a_1y_1^2 + a_2y_2^2 = b$.*

Proof. Q is universal in $\mathbb{F}_p$, i.e. it represents all b in $\mathbb{F}_p$, since we have

$$\#\{a_1x_1^2 | x_1 \in \mathbb{F}_p\} = \#\{b - a_2x_2^2 | x_2 \in \mathbb{F}_p\} = \tfrac{p+1}{2}$$
$$\Rightarrow \quad \{a_1x_1^2 | x_1 \in \mathbb{F}_p\} \cap \{b - a_2x_2^2 | x_2 \in \mathbb{F}_p\} \neq \emptyset$$
$$\Rightarrow \quad \exists \, \tilde{x}_1, \tilde{x}_2 \in \mathbb{F}_p : a_1\tilde{x}_1^2 = b - a_2\tilde{x}_2^2.$$

Let now $b \not\equiv 0 \, (\mathrm{mod} \, p)$ and without loss of generality $\tilde{x}_1 \not\equiv 0 \, (\mathrm{mod} \, p)$, then with $f(x) := a_1x^2 - (b - a_2\tilde{x}_2^2)$ it is

$$|f(\tilde{x}_1)|_p \leq \frac{1}{p} < |f'(\tilde{x}_1)|_p^2 = |2a_1\tilde{x}_1|_p^2 = 1.$$

Due to Hensel's Lemma there exists a $y_1 \in \mathbb{Z}_p$ such that $a_1y_1^2 + a_2\tilde{x}_2^2 = b$. $\square$

Proof of Proposition 1.3.5. As in the proof of Theorem 1.3.4 one can assume that Q is of the form

$$Q(x) = x_1^2 + \ldots + x_{n-1}^2 + \det(Q)x_n^2.$$

Due to Corollary 1.3.7 the form $x_1^2 + x_2^2$ represents all $b \equiv -\det(Q) \, (\mathrm{mod} \, p)$ in $\mathbb{Z}_p$. Hence for all m there exists an $x \in \mathbb{Z}_p^n$ such that $Q(x) = m$. $\square$

To treat the "critical" prime numbers p later on we need one more fact:

Proposition 1.3.8. *Let Q be an n-dimensional classically integral form, $n \geq 2$ and p a prime. Then Q represents a number m over $\mathbb{Z}_p$ if and only if Q represents m modulo p^{r+1}, where p^r is the highest power of p dividing $4m$.*

Proof. Let $u = r$ (if p is odd) or $u = r - 2$ (if $p = 2$), where p^u is the highest power of p dividing m and let $x \in \mathbb{Z}^n$ such that

$$x^T A x = m + p^{r+1} \cdot l,$$

where $p \nmid l$. Clearly $x_{r+1} := x$ is a representation in $\mathbb{Z}_p^n$ of m modulo p^{r+1}. Now we will use an induction argument and define the vector

$$x_{r+2} := x_{r+1} - \frac{1}{2} \cdot p^{r+1} x_{r+1} (x_{r+1}^T A x_{r+1})^{-1} \cdot l.$$

Since $x_{r+1} \in \mathbb{Z}_p^n$ and $r + 1 > u$ it follows that $x_{r+2} \in \mathbb{Z}_p^n$ too. Then we have (with $x_{r+1}^T A x_{r+1} = m + p^{r+1} \cdot l$)

$$
\begin{aligned}
x_{r+2}^T A x_{r+2} &= m + p^{r+1}l - \tfrac{1}{2}p^{r+1}l - \tfrac{1}{2}p^{r+1}l + \tfrac{1}{4}p^{2(r+1)}l^2(x_{r+1}^T A x_{r+1})^{-1} \\
&= m + \tfrac{1}{4}p^{2(r+1)}l^2(x_{r+1}^T A x_{r+1})^{-1}.
\end{aligned}
$$

Hence for some p-adic unit l' we get

$$
x_{r+2}^T A x_{r+2} = m + \frac{1}{4}p^{2(r+1)-u} \cdot l' = \begin{cases} m + p^{r+2} \cdot \frac{l'}{4}, & p \neq 2 \\ m + p^{r+2} \cdot l', & p = 2. \end{cases}
$$

Obviously the sequence x_{r+s} converges to a solution in $\mathbb{Z}_p^n$. $\qquad\square$

A major quantitative result along these lines was given by Siegel [Si] who was the first one to discover the concrete connection between local and global representability. For a detailed historical survey see [CoSl1]. His result allows us to think of our local information as a weighted average of information over the classes in the genus, cf. [Ca, Chap.9.6], [Ha, p.4]. We shall not have occasion to use it later, but it is hard to resist the temptation to praise Siegel's theorem at this place.

In a more general way we look at the representability of forms by forms, i.e. a form M with matrix B in n' variables is represented by an n-dimensional form Q with matrix A, $n > 1$ and $n' \leq n$, if there exists an $X \in \mathbb{Z}^{n \times n'}$ such that $X^T A X = B$. Furthermore we denote by $b_Q(M)$ the number of representations of M by Q and by $b_q(M)$ the number of representations of M by Q modulo $q = p^a$. If p^b is the highest power of p dividing $(2 \cdot \det(M))^2$ then the number

$$
\beta_p(M) := \begin{cases} q^{\frac{n'(n'+1)}{2} - nn'} \cdot b_q(M), & n > n' \\ \frac{1}{2}q^{-\frac{(n-1)n}{2}} \cdot b_q(M), & n = n' \end{cases}
$$

is independent of a for all $a > b$, cf. [Si, Hilfsatz 13]. We call these numbers the *local representation densities* of M.

Now we are able to state Siegel's result [Si, Hauptsatz]:

Theorem 1.3.9 (Siegel). *Let* gen(Q) *be the (disjoint) union of classes represented by forms Q_i and let $\varepsilon = 1$ if $n > n'+1$ or $n = n' = 1$ and $\varepsilon = \frac{1}{2}$ if $n = n'+1$ or $n = n' > 1$, then we have*

$$b_{\text{gen}(Q)}(M) := \frac{\sum_{Q_i} \frac{b_{Q_i}(M)}{b_{Q_i}(Q_i)}}{\sum_{Q_i} \frac{1}{b_{Q_i}(Q_i)}} = \varepsilon \cdot \beta_\infty(M) \prod_p \beta_p(M),$$

where β_∞ is a constant depending only on n, n', $\det(Q)$ and $\det(M)$ and can be seen as the presumable number of representations of M.

More precisely let $\mathcal{M}$ be a neighbourhood of M and let $\mathcal{X}$ be the set of matrices X such that $X^T A X \in \mathcal{M}$. Consider $\mathcal{M}$ and $\mathcal{X}$ as subsets of $\frac{n'(n'+1)}{2}$- and nn'-dimensional space respectively, then we define

$$\beta_\infty(M) := \lim_{\mathcal{M} \to M} \frac{\text{vol}(\mathcal{X})}{\text{vol}(\mathcal{M})}.$$

The value is

$$\beta_\infty(M) = \frac{\pi^{\frac{n'(2n-n'+1)}{4}}}{\Gamma(\frac{n-n'+1}{2})\Gamma(\frac{n-n'+2}{2})\dots\Gamma(\frac{n}{2})} \cdot \det(Q)^{-\frac{n'}{2}} \cdot \det(M)^{\frac{n-n'-1}{2}}$$

cf. [Si, Hilfssatz 26], here Γ is the *Gamma function* from analysis.

We have two interesting special cases.
First $Q = M$ gives a formula for the so called *weight of the genus* $W(Q)$ of the form Q:

$$W(Q) := \sum_{Q_i} \frac{1}{b_{Q_i}(Q_i)} = \frac{2\Gamma(\frac{1}{2})\Gamma(\frac{2}{2})\dots\Gamma(\frac{n}{2})\det(Q)^{\frac{n+1}{2}}}{\pi^{\frac{n(n+1)}{4}}\prod_p \beta_p(Q)}.$$

And secondly if $n' = 1$, so M is the number m, Siegel's formula reduces to

$$b_{\text{gen}(Q)}(m) = \varepsilon \cdot \frac{m^{\frac{n-2}{2}}\pi^{\frac{n}{2}}}{\Gamma(\frac{n}{2}) \cdot \det(Q)^{\frac{1}{2}}} \prod_p \beta_p(m).$$

One additional consequence is that Proposition 1.3.2 follows immediately since the infinite product over the $\beta_p(m)$ converges and all the $\beta_p(m) > 0$ if m is locally represented, cf. [Si, Hilfsatz 25].

1.4   Sphere packings

In Conjecture 2.1.1 lattices with the best sphere packings are involved. It is a classical problem cf. [CoSl4, Chap.1], to find out how densely a large number of identical balls can be packed together. So the density of a packing can be seen as proportion of the space that is occupied by these balls. More precisely, see also [NeXi]:

A *packing* in $\mathbb{R}^n$ is a set P of points such that the Euclidean distance

$$d(P) := \inf_{\substack{p_1, p_2 \in P \\ p_1 \neq p_2}} d(p_1, p_2)$$

is positive. Let $B_n(X)$ be the n-dimensional ball with radius $\sqrt{X}$. Now we build the union of all balls with radius $\frac{d(P)}{2}$ where the points of P are the centers: $U(P) := \{x \in \mathbb{R}^n : \exists p \in P, d(x,p) \leq \frac{d(P)}{2}\}$. Then the *density* of P is defined by

$$\Delta(P) := \limsup_{X \to \infty} \frac{\text{vol}(U(P) \cap B_n(X^2))}{\text{vol}(B_n(X^2))},$$

where as usual $\text{vol}(S)$ denotes the volume of a subset $S \subset \mathbb{R}^n$.

Hence for a lattice Γ the density is the proportion of the volume of one sphere (with radius $\frac{1}{2} \cdot \min_{\gamma \in \Gamma \setminus \{0\}} |\gamma|$) and the covolume of Γ:

$$\Delta(\Gamma) = \frac{\text{vol}(B_n(\frac{1}{4} \cdot \min_{x \in \mathbb{Z}^n \setminus \{0\}} Q_\Gamma(x)))}{\text{cov}(\Gamma)}.$$

1.5 Root lattices

Root lattices play a crucial role in many questions about lattices, so as in the question of the best sphere packing or the highest kissing number cf. Remark 3.2.2. By a *root lattice* we mean a lattice which is generated by a (reducible) root system. For more information about root systems one can refer to [Se2]. We use the same notation for the lattices as the usual one for the root systems. The irreducible root lattices occuring here are A_n, D_n, E_6, E_7 and E_8. These lattices can be characterised by Coxeter graphs cf. Table 1.1, where all n roots, respectively the basis vectors, have the same length and the angle between two roots is either $\frac{\pi}{3}$ or $\frac{\pi}{2}$ depending on whether they are connected or not.

Table 1.1: Root systems

1.6   Asymptotic notation

In the following we will characterise the limiting behaviour of real functions when the argument tends towards infinity with the common $\mathcal{O}$ and Ω symbols.

When a function is bounded above respectively dominated by another function asymptotically we use the upper and lower case $\mathcal{O}$.

Definition 1.6.1. Let f and g be two functions over the real or natural numbers.

(a) $f = \mathcal{O}(g) \quad :\Leftrightarrow \quad 0 \leq \limsup\limits_{x \to \infty} \left| \frac{f(x)}{g(x)} \right| < \infty$

$\qquad\qquad\qquad \Leftrightarrow \quad \exists\, c > 0 \; \exists\, x_0 \; \forall\, x > x_0 : |f(x)| \leq c \cdot |g(x)|$

(b) $f = o(g) \quad :\Leftrightarrow \quad \lim\limits_{x \to \infty} \left| \frac{f(x)}{g(x)} \right| = 0$

$\qquad\qquad\qquad \Leftrightarrow \quad \forall\, c > 0 \; \exists\, x_0 \; \forall\, x > x_0 : |f(x)| < c \cdot |g(x)|$

The Ω is used as negation of o. This means that a function exceeds the comparative function for infinitely many values. (In computer science, "infinitely many" often is replaced by "all large".)

Definition 1.6.2. Let f and g be two real functions, $g > 0$.

(a) $f = \Omega(g) \quad :\Leftrightarrow \quad f \neq o(g)$

$\qquad\qquad\qquad \Leftrightarrow \quad \limsup\limits_{x \to \infty} \left| \frac{f(x)}{g(x)} \right| > 0$

(b) $f = \Omega_+(g) \quad :\Leftrightarrow \quad \limsup\limits_{x \to \infty} \frac{f(x)}{g(x)} > 0$

(c) $f = \Omega_-(g) \quad :\Leftrightarrow \quad -f = \Omega_+(g)$

(d) $f = \Omega_\pm(g) \quad :\Leftrightarrow \quad f = \Omega_+(g) \wedge f = \Omega_-(g)$

Chapter 2

Conjecture of Schmutz Schaller

2.1 Introduction

Let Γ be a lattice in the Euclidean space $\mathbb{R}^n$. We sort the lengths of elements of Γ according to size:

$$0 = \gamma_0 < \gamma_1 < \gamma_2 < \gamma_3 < \gamma_4 < ...$$

where γ_k is called the *k-th length* of Γ. It is important that we leave out the multiplicities of these lengths.

Paul Schmutz Schaller formulated in [Schm2, p.201] the following conjecture, see also [Schm1]:

Conjecture 2.1.1 (Schmutz Schaller). *In dimensions 2 to 8 the lattices with the best lattice sphere packings have "maximal lengths", i.e. for all $k > 0$ their k-th length is strictly greater than the k-th length of any other lattice in the same dimension with the same covolume.*

This conjecture was motivated by considerations from hyperbolic geometry. Schmutz Schaller showed that in certain cases there exist surfaces, such that the lengths of their closed geodesics (without multiplicities) are maximal among all surfaces in their moduli space, cf. [Schm1, p.204].

Remark 2.1.2. Hence an alternative formulation of Conjecture 2.1.1 in terms of flat tori is that the tori corresponding to the special lattices mentioned above are extremal with respect to the k-th length of closed geodesics among the flat tori of the same dimension and volume.

The question of the densest sphere packing cf. Section 1.4 has a long history and is still unsolved for $n > 3$. In dimension $n = 1$ it is the trivial lattice $\mathbb{Z}$, for $n = 2$ the regular hexagonal lattice has the highest density. Gauß showed in [Ga] that the face-centered cubic lattice

$$\Sigma := \alpha \cdot \left(\mathbb{Z} \begin{pmatrix} 1 \\ 0 \\ 0 \end{pmatrix} \oplus \mathbb{Z} \begin{pmatrix} 1/2 \\ \sqrt{3}/2 \\ 0 \end{pmatrix} \oplus \mathbb{Z} \begin{pmatrix} 1/2 \\ \sqrt{3}/6 \\ \sqrt{6}/3 \end{pmatrix} \right) , \alpha := \sqrt[6]{2}$$

has the densest lattice packing. The general statement, that is known as Kepler's Conjecture, was proved by Thomas Hales in 1998. Unfortunately his computer-based proof has not yet been verified completely. Furthermore for $4 \leq n \leq 8$ the best (unique) lattice packings are known, cf. [CoSl4, Chap.1.1]. If we consider also nonlattice packings then the question is still open in these dimensions.

Obviously, for Conjecture 2.1.1 it is sufficient to know the best lattice packings. All these are irreducible root lattices. With the usual notation we have the following densest lattice packings in dimensions 1 to 8:

Table 2.1: Lattices with the best lattice sphere packings

n	1	2	3	4	5	6	7	8
$2\widetilde{\Sigma}_n$	A_1	A_2	A_3	D_4	D_5	E_6	E_7	E_8

Since in higher dimensions we do not even know the best lattice packings the conjecture would not make sense for $n \geq 9$.

Kühnlein [Kü1, Thm.1] showed the conjecture in dimension 2 for lattices with bounded multiplicities, cf. Section 2.5. For arithmetic lattices so far only examples are known, for instance Moree and te Riele [MoRi, Thm.1] proved that the lengths of the regular hexagonal lattice dominate the lengths of the regular square lattice.

We will see in Section 2.2 that the conjecture is asymptotically true because the lattices with the best known sphere packings minimize the so called Erdös number. But as a result of the investigation of the 3-dimensional lattice with the second smallest Erdös number we will see in Section 2.3 that the conjecture is not true for ternary lattices. However, it seems that there is only one exception: one lattice, where for one length the conjecture fails.

2.2   Asymptotics

Theorem 2.2.1. *For $2 \leq n \leq 8$ let Σ_n be the normalised (covolume 1) n-dimensional lattice with the best lattice sphere packing and Γ_n be any normalised lattice in $\mathbb{R}^n$ not congruent to Σ_n. Then the k-th length of Σ_n is strictly greater than the k-th length of Γ_n if k is big enough.*

Proof. The *Erdös number* of an n-dimensional lattice Γ is given by

$$E_\Gamma := F_\Gamma \cdot \mathrm{cov}(\Gamma)^{\frac{1}{n}},$$

where F_Γ for $n \geq 3$ is given by

$$F_\Gamma := \lim_{X \to \infty} \frac{P_\Gamma(X)}{X}.$$

Here P_Γ is the *population function* of the associated form, i.e. the function that counts the number of values of the form, that do not exceed X:

$$
\begin{aligned}
P_\Gamma(X) \;:=\; & \# \{x < X : \exists v \in \mathbf{Z}^n, Q_\Gamma(v) = x\} \\
=\; & \# \{x < X : \exists \gamma \in \Gamma, |\gamma|^2 = x\} .
\end{aligned}
$$

For $n = 2$ this definition must be modified: $F_\Gamma := \lim_{X \to \infty} \frac{P_\Gamma(X)}{X} \sqrt{\log X}$.

Conway and Sloane [CoSl2] showed that the lattices with minimal Erdös number in dimensions $3 \leq n \leq 8$ are (up to a scaling factor) precisely the integral lattices with minimal covolume. And those are the lattices with the best lattice sphere packings. The analogous result for binary lattices was obtained in [MoOs, Thm.1].

Now let Γ_n be any lattice with covolume 1 not congruent to Σ_n. In case of $E_\Gamma = \infty$ the claim is even true for lattices with arbitrary covolume. (In particular the Erdös number of any non-arithmetic lattice is infinite for $n \geq 3$.) In case $E_\Gamma = F = \lim_{X \to \infty} \frac{P_\Gamma(X)}{X} < \infty$ we have

$$P_\Gamma(X) = E_\Gamma X + o(X) \text{ and } P_\Sigma(X) = E_\Sigma X + o(X).$$

Since E_Σ is minimal, i.e. $E_\Sigma < E_\Gamma$, the number of lengths of Γ_n less than $\sqrt{X}$ grows asymptotically faster than the analogous number for Σ_n. Hence $\lambda_k < \sigma_k$ if k is big enough. $\qquad\square$

Table 2.2: The four ternary lattices with the smallest Erdös numbers

Lattice	A_3	$A_2 \times A_1$	$A_2 \times 3A_1$	$\mathbf{Z}^3$
Erdös number	$\frac{11}{24}\sqrt[3]{4}$	$\frac{7}{16}\sqrt[3]{6}$	$\frac{5}{16}\sqrt[3]{18}$	$\frac{5}{6}$

For $n = 3$ the minimal Erdös number is $E_\Sigma = \frac{11}{24}\sqrt[3]{4}$. In [CoSl2, p.86] the ternary lattices with the next smallest Erdös numbers are given as well, cf. Table 2.2. This means that the reducible root lattice $A_2 \times A_1$ asymptotically dominates all other lattices not congruent to Σ. Therefore it is interesting to check Conjecture 2.1.1 for this lattice.

2.3 A counter-example

2.3.1 The face-centered cubic lattice

With the information of Section 1.3 we are able to give the numbers that are represented by the form Q_Σ. Therefore we investigate the even values of the form $Q_{A_3} = Q_{2\widetilde{\Sigma}}$, where $\widetilde{\Sigma} := \frac{1}{\alpha}\Sigma$.

$$Q_{2\widetilde{\Sigma}}(x) = x^T \cdot \begin{pmatrix} 2 & 1 & 1 \\ 1 & 2 & 1 \\ 1 & 1 & 2 \end{pmatrix} \cdot x = 2 \cdot (x_1^2 + x_2^2 + x_3^2 + x_1 x_2 + x_1 x_3 + x_2 x_3)$$

As mentioned in the proof of Theorem 2.2.1, $Q_{2\widetilde{\Sigma}}$ is up to equivalence the unique classically integral form with minimal determinant. Hence a form that is locally equivalent to $Q_{2\widetilde{\Sigma}}$ is also (globally) equivalent: $\Rightarrow \text{gen}(Q_{2\widetilde{\Sigma}})$ has only one class.

Due to Proposition 1.3.5 there exist no congruence conditions for primes $p \neq 2$. With Proposition 1.3.8 it remains to check the following cases:

(a) If $m \equiv 2\,(\text{mod}\,8)$ or $m \equiv 6\,(\text{mod}\,8)$ there exists an $x \in \mathbb{Z}^3$ such that $Q_{2\widetilde{\Sigma}}(x) \equiv m\,(\text{mod}\,16)$, because $Q_{\widetilde{\Sigma}}$ represents $1, 3, 5, 7$.

(b) If $m \equiv 4\,(\text{mod}\,8)$, then $Q_{2\widetilde{\Sigma}}(x) \equiv m\,(\text{mod}\,32)$ is solvable if and only if $m \not\equiv 28\,(\text{mod}\,32)$, because $Q_{\widetilde{\Sigma}}$ represents $2, 6, 10$ but not 14 (so 14 cannot be represented locally).

(c) If $m \equiv 0\,(\text{mod}\,8)$ it is easy to see that x_1, x_2 and x_3 have to be even.
$\Rightarrow Q_{2\widetilde{\Sigma}}(x) = 4 \cdot Q_{2\widetilde{\Sigma}}(\frac{x}{2})$
$\Rightarrow Q_{2\widetilde{\Sigma}}$ represents m if and only if $Q_{2\widetilde{\Sigma}}$ represents $\frac{m}{4}$.

Thus we have shown that $Q_{2\widetilde{\Sigma}}$ represents all positive even integers except those of the form

$$4^a(32t + 28), \quad a, t \in \mathbb{N}_0$$

and represents none of this form. Hence we have:

Lemma 2.3.1. $Q_{\widetilde{\Sigma}}$ *represents a positive integer m if and only if*

$$m \neq 4^a(16t + 14) \text{ for all } a, t \in \mathbb{N}_0.$$

But this lattice is very particular, so it would be more elegant if we exploit that the lattice $2\widetilde{\Sigma} = A_3$ is the even sublattice of $\mathbb{Z}^3$ whose coordinates add up to an even number; $2\widetilde{\Sigma} = \{x \in \mathbb{Z}^3 : x_1 + x_2 + x_3 \equiv 0\,(\text{mod}\,2)\} = \{x \in \mathbb{Z}^3 : |x|^2 \equiv 0\,(\text{mod}\,2)\}$. For the cubic lattice it is well known that the associated quadratic form represents all numbers unequal to $4^a(8t + 7)$. Therefore $Q_{2\widetilde{\Sigma}}$ represents all even numbers that are unequal to $4^a(8t + 7)$, for all $a \in \mathbb{N}$ and $t \in \mathbb{N}_0$.

Now we can estimate the k-th length σ_k of Σ from below:

Proposition 2.3.2. *For the k-th length σ_k of Σ ($k \geq 4$) we have:*

$$\sigma_k \geq \alpha \cdot \sqrt{\tfrac{17}{16}k - \tfrac{14}{16}}.$$

Proof. For the k-th length of $\widetilde{\Sigma}$ we have: $\widetilde{\sigma}_k^2 \geq k + \#\{t \in \mathbb{N}_0 : 16t + 14 \leq k\}$.

$$\Rightarrow \sigma_k^2 \geq \alpha^2 \cdot (k + \underbrace{\#\{t \in \mathbb{N}_0 : 16t + 14 \leq k\}}_{=\left[\frac{k-14}{16}+1\right] \geq \frac{k}{16} - \frac{14}{16}}) \geq \alpha^2 \cdot (\tfrac{17}{16}k - \tfrac{14}{16})$$

$\square$

2.3.2 The honeycomb lattice

The primitive honeycomb lattice

$$\Lambda := \beta \cdot \left(\mathbb{Z} \begin{pmatrix} 1 \\ 0 \\ 0 \end{pmatrix} \oplus \mathbb{Z} \begin{pmatrix} 1/2 \\ \sqrt{3}/2 \\ 0 \end{pmatrix} \oplus \mathbb{Z} \begin{pmatrix} 0 \\ 0 \\ 1 \end{pmatrix} \right), \; \beta := \sqrt[6]{\tfrac{4}{3}}$$

has the second smallest Erdös number cf. Chapter 2.2. It is also highly symmetric and has a small covolume or big normalisation factor β.

Since $\frac{1}{2}A_2 \subset \widetilde{\Lambda} := \frac{1}{\beta}\Lambda$ and $\mathbb{Z}^2 \subset \widetilde{\Lambda}$, the associated integral form

$$Q_{\widetilde{\Lambda}}(x) := x^T \cdot \begin{pmatrix} 1 & \frac{1}{2} & 0 \\ \frac{1}{2} & 1 & 0 \\ 0 & 0 & 1 \end{pmatrix} \cdot x = x_1^2 + x_2^2 + x_3^2 + x_1 x_2$$

represents all numbers m which are represented by $Q_{\frac{1}{2}A_2}$ and $Q_{\mathbb{Z}^2}$. Hence $Q_{\widetilde{\Lambda}}$ certainly represents those numbers whose prime factors $p \equiv 2 \,(\mathrm{mod}\ 3)$ all divide the number in an even power and those numbers whose prime factors $p \equiv 3 \,(\mathrm{mod}\ 4)$ all divide the number in an even power. Therefore the first possible exception is 6; it turns out that in fact 6 is not represented. Hence the sixth length of Λ (with 0 as zeroth length) is

$$\lambda_6 = \beta \cdot \sqrt{7} > 2.77 > 2.75 > \alpha \cdot \sqrt{6} = \sigma_6.$$

We see here that Conjecture 2.1.1 does not hold for dimension 3. Now the question arises how many such exceptional lengths λ_k there are in Λ.

We could study the congruence conditions as with Σ but it would be more elegant if we exploit that $Q_{\widetilde{\Lambda}}$ and $Q_{\Delta_3} := x_1^2 + 3x_2^2 + x_3^2$ have the same set of lengths. With the same basis as in the definition of Λ the lattices $\Delta_1 := \mathbb{Z}b_1 \oplus \mathbb{Z}2b_2 \oplus \mathbb{Z}b_3$, $\Delta_2 := \mathbb{Z}2b_1 \oplus \mathbb{Z}b_2 \oplus \mathbb{Z}b_3$ and $\Delta_3 := \mathbb{Z}(-b_1 + b_2) \oplus \mathbb{Z}(b_1 + b_2) \oplus \mathbb{Z}b_3$ are sublattices of $\widetilde{\Lambda}$ of index 2. Since the union of the three sublattices is $\widetilde{\Lambda}$ and the associated quadratic forms Q_{Δ_1}, Q_{Δ_2} and Q_{Δ_3} are all in the same class the three lattices have all the same length spectrum, hence the same length spectrum as that of $\widetilde{\Lambda}$ itself.

It is well known which numbers are represented by the form $Q_{\Delta_3} =: Q_\Delta$. Already Ramanujan [Ram, p.13] knew the conditions, the formal proof of his result is given in [Di1, Thm.III]:

Lemma 2.3.3. Q_Δ (and so $Q_{\tilde{\Lambda}}$) represents a positive integer m if and only if

$$m \neq 9^a(9t + 6) \text{ for all } a, t \in \mathbb{N}_0.$$

We prove this lemma with a short calculation:

Proof. First we show that every form Q in the genus of Q_Δ is equivalent to Q_Δ. Q has to be equivalent to a reduced form with determinant 3. By reference to a table [Ei, p.169] there are only two such forms $Q_1(x) = x_1^2 + x_2^2 + 3x_3^2$ and $Q_2(x) = x_1^2 + 2x_2^2 + 2x_3^2 - x_2x_3$. Since Q_2 represents 6 it cannot be locally equivalent to Q_Δ. Hence Q is equivalent to Q_1, which is in turn equivalent to Q_Δ:

$$\Rightarrow \text{gen}(Q_\Delta) \text{ has only one class.}$$

Once again due to Proposition 1.3.5 we have only to consider the primes 2 and 3.

(a) $p = 2$: For $m \equiv 1 \,(\text{mod } 2)$ or $\equiv 2 \,(\text{mod } 4)$ exists an $x \in \mathbb{Z}^n$ with $Q_\Delta(x) \equiv m \,(\text{mod } 2^3)$ or $Q_\Delta(x) \equiv m \,(\text{mod } 2^4)$, because Q_Δ represents $1, 3, 5$ and $2, 22, 10, 14$. If $m \equiv 0 \,(\text{mod } 4)$ we can write $m = 4^k t$, where $t \not\equiv 0 \,(\text{mod } 4)$. So there exists an x_0 with $Q_\Delta(x_0) \equiv t \,(\text{mod } 2^3)$ or $Q_\Delta(x_0) \equiv (\text{mod } 2^4)$ depending on t is odd or even. Hence $Q_\Delta(2^k x_0) \equiv 4^k t \,(\text{mod } 2^{2k+3})$ or $(\text{mod } 2^{2k+4})$. With Proposition 1.3.8 it follows that Q_Δ is universal in $\mathbb{Z}_2$.

(b) $p = 3$: For $m \equiv 1 \,(\text{mod } 3)$, $m \equiv 2 \,(\text{mod } 3)$ or $m \equiv 3 \,(\text{mod } 9)$ there exists an x with $Q_\Delta(x) \equiv m \,(\text{mod } 3)$ or $Q_\Delta(x) \equiv m \,(\text{mod } 3^2)$, because Q_Δ represents $1, 2$ and 3. If $m \equiv 0 \,(\text{mod } 9)$ we see that $x_1, x_2, x_3 \equiv 0 \,(\text{mod } 3)$. So Q_Δ represents m if and only if Q_Δ represents $\frac{m}{9}$. Since Q_Δ does not represent 6 (and so it cannot represent 6 locally) it follows that $Q_\Delta(x) \neq 6 \,(\text{mod } 3^2)$.

$\square$

Now we can estimate the length λ_k from above.

Proposition 2.3.4. *For the k-th length λ_k of Λ we have:* $\lambda_k \leq \beta \cdot \sqrt{\frac{840}{727} k}$.

Proof. If $9^a \cdot 6 > k$ for $a \in \mathbb{N}_0$ we have $\#\{t \in \mathbb{N}_0 : 9^a(9t + 6) \leq k\} = 0$, hence

$$\#\left\{(a, t) \in \mathbb{N}_0^2 : 9^a(9t + 6) \leq k\right\}$$

$$= \sum_{a=0}^{\left[\log_9\left(\frac{k}{6}\right)\right]} \#\{t \in \mathbb{N}_0 : 9^a(9t + 6) \leq k\}$$

$$= \sum_{a=0}^{\left[\log_9\left(\frac{k}{6}\right)\right]} \left[\frac{k - 9^a \cdot 6}{9^a \cdot 9} + 1\right] \leq \sum_{a=0}^{\left[\log_9\left(\frac{k}{6}\right)\right]} \left(\frac{k}{9^a \cdot 9} + \frac{1}{3}\right)$$

$$\leq \frac{k}{9} \cdot \underbrace{\sum_{a=0}^{\infty} \frac{1}{9^a}}_{= \frac{1}{1 - \frac{1}{9}}} + \sum_{a=0}^{\left[\log_9\left(\frac{k}{6}\right)\right]} \frac{1}{3} = \frac{1}{8}k + \frac{1}{3\ln(9)}\ln(k) - \frac{\ln(6)}{3\ln(9)} .$$

It is easy to see that $\frac{1}{\ln(9)}\ln(k) - \frac{\ln(6)}{\ln(9)} \le \frac{1}{35}k$ for all $k \in \mathbb{N}$ and this estimate will do all that we need:

$$\#\{(a,t) \in \mathbb{N}_0^2 : 9^a(9t+6) \le k\} \le \tfrac{1}{8}k + \tfrac{1}{3} \cdot \tfrac{1}{35}k = \tfrac{113}{840}k \; (*)$$

$$\Rightarrow \tilde{\lambda}_k^2 \le k + \#\{(a,t) \in \mathbb{N}_0^2 : 9^a(9t+6) \le k + \tfrac{113}{840}k \sum_{j=0}^{\infty} \left(\tfrac{113}{840}\right)^j\}$$

$$\le k + \#\{(a,t) \in \mathbb{N}_0^2 : 9^a(9t+6) \le \tfrac{840}{727}k\} \underset{(*)}{\le} \tfrac{840}{727}k.$$

$\square$

The lower bound of σ_k of Propsition 2.3.2 dominates the upper bound of λ_k of Proposition 2.3.4 for $k \ge 17$. If we compare the first 16 lengths as well, cf. Table 2.3, we get:

Proposition 2.3.5. *Let Σ and Λ be defined as above and $k \in \mathbb{N}$, then $\sigma_k > \lambda_k$ if and only if $k \ne 6$.*

Table 2.3: Lengths 1-16

k	1	2	3	4	5	6	7	8
$\tilde{\sigma}_k^2$	1	2	3	4	5	6	7	8
σ_k^2	1.26	2.5	3.8	5.0	6.3	**7.56**	8.82	10.1
λ_k^2	1.1	2.2	3.3	4.4	5.5	**7.70**	8.81	9.9
$\tilde{\lambda}_k^2$	1	2	3	4	5	7	8	9

k	9	10	11	12	13	14	15	16
$\tilde{\sigma}_k^2$	9	10	11	12	13	14	15	16
σ_k^2	11.3	12.6	13.9	15.1	16.4	18.9	20.2	21.4
λ_k^2	11.0	12.1	13.2	14.3	15.4	17.6	18.7	19.8
$\tilde{\lambda}_k^2$	10	11	12	13	14	16	17	18

In some respect the estimates in Proposition 2.3.2 and 2.3.4 are unnecessarily weak. Instead of estimating the k-th length directly we can also use the population function from Chapter 2.2 to get sharper bounds:

Proposition 2.3.6. *For $X > 0$ we have: $P_\Sigma(X) \le \frac{1}{\alpha^2}\frac{15}{16}X + \frac{14}{16}.$*

Proof. We have

$$P_\Sigma(X) = \#\{x < X : \exists v \in \mathbb{Z}^3, Q_{\widetilde{\Sigma}}(v) = \tfrac{1}{\alpha^2}x\}$$
$$= \#\{x < \tfrac{1}{\alpha^2}X : \exists v \in \mathbb{Z}^3, Q_{\widetilde{\Sigma}}(v) = x\} = P_{\widetilde{\Sigma}}(\tfrac{1}{\alpha^2}X)$$

and

$$P_{\widetilde{\Sigma}}(X) = \#\{x < X : x \in \mathbb{N}_0 \wedge x \neq 4^a(16t+14), (a,t) \in \mathbb{N}_0^2\}$$
$$\leq \#\{x < X : x \in \mathbb{N}_0 \wedge x \neq (16t+14), t \in \mathbb{N}_0\}$$
$$\leq X - \left[\tfrac{X-14}{16} + 1\right] \leq \tfrac{15}{16}X + \tfrac{14}{16}.$$

$\square$

Proposition 2.3.7. *For $X > 0$ we have:*

$$P_\Lambda(X) \geq \tfrac{1}{\beta^2}\tfrac{7}{8}X - \tfrac{1}{3\ln(9)}\ln(\tfrac{1}{\beta^2}X) + \tfrac{\ln(6)}{3\ln(9)} - 1.$$

Proof. As in Proposition 2.3.6 we have

$$P_{\widetilde{\Lambda}}(X) = \#\{x < X : x \in \mathbb{N}_0 \wedge x \neq 9^a(9t+6), (a,t) \in \mathbb{N}_0^2\}$$
$$\geq [X] - \#\{(a,t) \in \mathbb{N}_0^2 : 9^a(9t+6) \leq X\}$$
$$\geq (X-1) - \tfrac{1}{8}X - \tfrac{1}{3\ln(9)}\ln(X) + \tfrac{\ln(6)}{3\ln(9)}.$$

$\square$

If for some $X_0 > 0$ the inequality $P_\Lambda(X_0) > P_\Sigma(X_0)$ holds, then

$$\sigma_{P_\Sigma(X)} > \lambda_{P_\Lambda(X)} > \lambda_{P_\Sigma(X)} \text{ for all } X \geq X_0.$$

Hence $\lambda_k < \sigma_k$ for all $k \geq P_\Sigma(X_0)$.
For $X_0 = 43$ we have $P_\Lambda(43) \geq 32.90$ and $P_\Sigma(43) \leq 32.87$, and so a dominance of the 32-nd length. We can see that for our purpose taking the estimate of the k-th length is more effective.

2.3.3 Other counter-examples?

With our simple strategy to look at "interesting" lattices, i.e. lattices with a small Erdös number, small covolume, high multiplicities etc., we were not able to find other counter-examples.

Reasoning as in Section 2.2 we find that the ternary lattices with the next smallest Erdös numbers do not contradict Conjecture 2.1.1. For example the form associated with $A_2 \times 3A_1$ has the same values as the form $x_1^2 + 3x_2^2 + 3x_3^2$. Hence it represents an integer m if and only if $m \neq 9^a(3t+2)$ cf. [Di1, Thm.IV]. With an analogue estimate we get an upper bound of the k-th length that is dominated by the lower bound of σ_k for $k \geq 20$. And for the first 19 lengths, σ_k is strictly greater; though sometimes it is really "close", e.g. $\sigma_2 \approx 1.587$ and $(a_2 \times 3a_1)_2 \approx 1.513$.

For the cubic lattice the associated quadratic form represents all numbers $m \neq 4^a(8t+7)$ and we can see once again that the σ_k are dominant for all k. The differences between the first lengths are clearly greater; since the factor α becomes larger and larger it would not be promising to go further in this direction with the next smallest Erdös numbers.

In higher dimensions this method would not be successful either, because all "interesting" forms become *universal*, i.e. they represent all positive integers. Therefore the statement becomes trivial because the lattices with the best known sphere packings have minimal covolume and so the greatest normalization factor. In dimension 4 the lattices with the first five smallest Erdös numbers are D_4, A_4, $A_3 \times A_1$, $A_2 \times A_2$ and $A_2 \times A_1 \times A_1$ cf. [CoSl2, p.87]. For $n \geq 3$, D_n are the even sublattices of $\mathbb{Z}^n$ whose coordinates add up to an even number (cf. Lemma 2.3.1, $D_3 = A_3$), hence $D_n = \{x \in \mathbb{Z}^3 : x_1 + ... + x_n \equiv 0 \,(\mathrm{mod}\,2)\} = \{x \in \mathbb{Z}^n : |x|^2 \equiv 0 \,(\mathrm{mod}\,2)\}$. And it is well known that every number can be written as a sum of four integer squares. Therefore Q_{D_4} represents all even numbers and so $Q_{\frac{1}{2}D_4}$ is universal. The universality of the form associated with A_4 can be seen by an easy local calculation, because 14, the first exception of $Q_{\overline{5}}$, is represented cf. proof of Lemma 2.3.1. For the other forms we can use that $Q_{\frac{1}{2}A_2}$ has the same lengths as the form $x_1^2 + 3x_2^2$ and $Q_{\frac{1}{2}A_3}$ as the form $x_1^2 + x_2^2 + 2x_3^2$. Hence we know a quaternary diagonal form with the same lengths. The universality of this diagonal form can be checked simply by a table [Di2] or with use of the remarkable *Conway-Schneeberger Fifteen Theorem*, which says that a classically integral form is universal if and only if it represents all positive integers up to 15 cf. [Bh, Thm.1].

For the binary conjecture the only "interesting" root lattice is the square lattice. For this lattice Moree and te Riele [MoRi, Thm.1] have shown (by completely different methods) the stronger statement, that the k-th length of the non-normalised lattice $\frac{1}{2}A_2$ is greater than or equal to the k-th length of the square lattice.

Finally we can resume that so far λ_6 is the only known length that contradicts the conjecture.

2.4 Irregular ternary forms

2.4.1 Introduction

The results of Section 2.3 may tempt us to believe that there are similar simple results for all ternary arithmetic forms, cf. Remark 2.5.3. It appears, however, to be beyond the reach of present methods to deal with Conjecture 2.1.1 in general because the situation is very different for forms belonging to a genus with more than one class, or more precisely for forms missing some integers which are not prohibited by congruence conditions. We call such a form *irregular*. There is no known effective way of determining the integers which are represented by irregular forms. In the following we discuss Conjecture 2.1.1 for two well-studied examples.

2.4.2   Ramanujan's form

In [Ram, p.14] Ramanujan investigated the representation of integers by quadratic forms. In a footnote he discussed among others the irregular ternary form

$$Q_{\tilde{P}}(x) := x_1^2 + x_2^2 + 10x_3^2 :$$

"Again, the even numbers which are not of the form $x^2 + y^2 + 10z^2$ are the numbers

$$4^\lambda(16\mu + 6),$$

while the odd numbers that are not of that form, viz.

$$3, 7, 21, 31, 33, 43, 67, 79, 87, 133, 217, 219, 223, 253, 307, 391, \ldots$$

do not seem to obey any simple law."

This form is often referred to in the literature as *Ramanujan's form*. The genus of $Q_{\tilde{P}}$ consists of two classes and

$$Q_{\tilde{P}'}(x) := 2x_1^2 + 2x_2^2 + 3x_3^2 - 2x_1 x_3$$

represents the other class. As a consequence of [DuSP, p.56 Cor.] the set of odd exceptions is finite, but so far an explicit bound is not known. Besides the elements in Ramanujan's list there are two more discovered exceptions: 679 and 2719. Ono and Soundararajan proved in [OnSo] depending on the validity of the generalized Riemann Hypotheses that these are actually all exceptions.

However, since the determinant of $Q_{\tilde{P}}$ is "great enough" we do not need the GRH to check Conjecture 2.1.1. Similar to Section 2.3.2 we have:

$$\# \left\{ (a, t) \in \mathbb{N}_0^2 : 4^a(16t + 6) \leq k \right\} \leq \frac{59}{528} k. \quad (*)$$

If we leave out every odd number ≥ 3 we get the estimate:

$$\tilde{\rho}_k^2 \leq (2k - 1) + \#\{(a, t) \in \mathbb{N}_0^2 : 4^a(16t + 6) \leq 2(k + \tfrac{59}{528} k \sum_{j=0}^{\infty} (\tfrac{59}{528})^j)\}$$

$$\underset{(*)}{\leq} 2\tfrac{118}{469} k - 1.$$

The lower bound of σ_k^2 of Proposition 2.3.2 dominates the upper bound of $\rho_k^2 := \frac{1}{\sqrt[3]{10}} \cdot \tilde{\rho}_k^2$ for $k \geq 3$, hence $\sigma_k > \rho_k$ for all k.

2.4.3   The first nontrivial genus

Again we saw in Section 2.4.2 that the determinant of a form is crucial for Conjecture 2.1.1. Therefore we investigate the first irregular ternary example, meaning that the determinant is smallest, the form

$$Q_{\tilde{K}} := x_1^2 + x_2^2 + 7x_3^2$$

or $Q_{\widetilde{K}'} := x_1^2 + 2x_2^2 + 4x_3^2 + 2x_1x_3$ which represents the second class respectively. Despite a long history this first nontrivial case is still not completely understood, cf. [Ka]. To deal with Conjecture 2.1.1 we first need the congruence conditions:

Lemma 2.4.1. *The numbers which are not represented by one form in* $\text{gen}(Q_{\widetilde{K}})$ *are those belonging to one of the three classes*

$$49^a(49t + 21), \quad 49^a(49t + 35) \quad or \quad 49^a(49t + 42) \ for \ a, t \in \mathbb{N}_0.$$

Proof. Due to Theorem 1.3.2 it is sufficient to check if an integer m is locally represented by one form in $\text{gen}(Q_{\widetilde{K}})$. Again with Proposition 1.3.5 we have only to consider the primes 2 and 7.

(a) $p = 2$: For $m \equiv 1 \, (\text{mod } 2)$ there exists an x with $Q_{\widetilde{K}}(x) \equiv m \, (\text{mod } 2^3)$, because $Q_{\widetilde{K}}$ represents $1, 11, 5, 7$. If $m \equiv 2 \, (\text{mod } 4)$ there exists an x with $Q_{\widetilde{K}}(x) \equiv m \, (\text{mod } 2^4)$, because $Q_{\widetilde{K}}$ represents $2, 38, 10, 30$. If $m \equiv 0 \, (\text{mod } 4)$ the proof is similar to the one of Lemma 2.3.3, hence $Q_{\widetilde{K}}$ is universal in $\mathbb{Z}_2$.

(b) $p = 7$: For $m \not\equiv 0 \, (\text{mod } 7)$ there exists an x with $Q_{\widetilde{K}}(x) \equiv m \, (\text{mod } 7)$, because $Q_{\widetilde{K}}$ represents $1, 2, 10, 4, 5, 13$. If $m \equiv 7 \, (\text{mod } 49)$, $m \equiv 14 \, (\text{mod } 49)$ or $m \equiv 28 \, (\text{mod } 49)$ there exists an x with $Q_{\widetilde{K}}(x) \equiv m \, (\text{mod } 49)$, because Q_K represents $7, 63, 28$. Since $\text{gen}(Q_{\widetilde{K}})$ does not represent $21, 35$ and 42 these numbers cannot be represented locally by any form in the genus: $Q_{\widetilde{K}}(x) \not\equiv 21 \, (\text{mod } 7^2)$, $Q_{\widetilde{K}}(x) \not\equiv 35 \, (\text{mod } 7^2)$ and $Q_{\widetilde{K}}(x) \not\equiv 42 \, (\text{mod } 7^2)$. If $m \equiv 0 \, (\text{mod } 49)$ we see that $x_1^2 + x_2^2 \equiv 0 \, (\text{mod } 7)$. That is if and only if $x_1 \equiv x_2 \, (\text{mod } 7) \equiv 0 \, (\text{mod } 7)$. So $x_1^2 + x_2^2 \equiv 0 \, (\text{mod } 49)$ and finally $x_3 \equiv 0 \, (\text{mod } 7)$. Hence $\text{gen}(Q_{\widetilde{K}})$ represents m if and only if it represents $\frac{m}{49}$.

$\square$

Again the additional exceptions do not seem to obey a simple law:

$$3, 6, 14, 19, 22, 31, 51, 55, 66, 94, 139, 142, 147, 154, 159, 166 \ldots$$

Yet with the following proposition from Kaplansky [Ka, Thm.2, Thm.3] we are able to count out "enough" numbers which are represented by $Q_{\widetilde{K}}$.

Proposition 2.4.2 (Kaplansky). *Let m be a number which is represented by $\text{gen}(Q_{\widetilde{K}})$. Then we have:*

(a) $Q_{\widetilde{K}}$ *represents* m *if* $m \equiv 0$ *or* $1 \, (\text{mod } 4)$.

(b) $Q_{\widetilde{K}}$ *represents* m *if* $m \equiv 2 \, (\text{mod } 3)$ *and* $m \neq 14t^2$.

Proposition 2.4.3. *For every number m which is represented by $Q_{\widetilde{K}}$ there exist at most $\left[\frac{m}{2}\right]$ exceptions less than m.*

Proof. Obviously if m is any number greater than 4 there exist at most $\left[\frac{m}{2}\right]$ numbers less than m which are congruent to 2 or 3 (mod 4). Since 2 is represented by $Q_{\tilde{K}}$ there exist at most $\left[\frac{m}{2}\right]$ such numbers for any represented m. Now we show that for every congruence exception less than m there exists a (different) smaller number r which is represented by $\text{gen}(Q_{\tilde{K}})$ such that $r \equiv 2 \,(\text{mod } 3)$, $r \neq 14t^2$ and $r \not\equiv 0, 1 \,(\text{mod } 4)$.

Let r'' be of the form $49^a(49t + 21)$, then $r'' \equiv t \,(\text{mod } 3)$.
Define $r' := r'' - 1, 2$ or 3 according to $t \equiv 0, 1$ or $2 \,(\text{mod } 3)$. Additionally we define

(a) $r := r'$ if $r' \not\equiv 0$ or $1 \,(\text{mod } 4)$,

(b) $r := r' - 3$ if $r' \equiv 1 \,(\text{mod } 4)$,

(c) $r := r' - 6$ if $r' \equiv 0 \,(\text{mod } 4)$ and $t \not\equiv 0 \,(\text{mod } 3)$,

(d) $r := r' - 9$ if $r' \equiv 0 \,(\text{mod } 4)$ and $t \equiv 0 \,(\text{mod } 3)$.

Hence $r \equiv 2 \,(\text{mod } 3)$, $r \not\equiv 0, 1 \,(\text{mod } 4)$ and $r \not\equiv 0 \,(\text{mod } 7)$, with Lemma 2.4.1 and Proposition 2.4.2 follows that $Q_{\tilde{K}}$ represents r with $0 < r < r''$.

We can do this analogously with the numbers of the form $r_2'' = 49^a(49t + 35)$ and $r_3'' = 49^a(49t + 42)$ but must be careful not to count a number twice. Since the numbers are congruent to $21, 35$ or $42 \,(\text{mod } 49)$ it is possible that $r_3'' - 8, 9, 10$ is equal to $r_2'' - 1, 2, 3$. Therefore we define

(a) $r_2 := r_2' - 9$ if $r_2' \equiv 3 \,(\text{mod } 4)$ and

(b) $r_2 := r_2' - 3$ if $r_2' \equiv 2 \,(\text{mod } 4)$.

The remaining cases are similar as above.
Hence we get for all congruence exceptions different (smaller) numbers which are represented and not congruent to 0 or $1 \,(\text{mod } 4)$. $\qquad\square$

Corollary 2.4.4. *The k-th length of Q_Σ is greater than the k-th length of Q_K for all $k \in \mathbb{N}$.*

Proof. Suppose $\sigma_k^2 = \alpha^2 \cdot \tilde{\sigma}_k^2 \leq \frac{1}{\sqrt[3]{7}} \cdot \tilde{\kappa}_k^2 = \kappa_k^2$ for one $k \in \mathbb{N}$.

$$\Rightarrow \tilde{\kappa}_k^2 \geq \alpha^2 \sqrt[3]{7} \cdot \tilde{\sigma}_k^2 \geq \alpha^2 \sqrt[3]{7} \cdot (\tfrac{17}{16}k - \tfrac{14}{16}) > 2.56k - 2.11$$

If $k \leq 3$ we have $\sigma_k > \kappa_k$, so let be $k \geq 4$:

$$\Rightarrow \tilde{\kappa}_k^2 > 2k$$

Hence $\tilde{\kappa}_k^2$ is of the form $2k + m$ with $k, m \in \mathbb{N}$ and there exist $k + m$ exceptions less than $\tilde{\kappa}_k^2$. Since

$$k + m > \tfrac{2k+m}{2} \geq \left[\tfrac{\tilde{\kappa}_k^2}{2}\right]$$

we have a contradiction to Proposition 2.4.3. $\qquad\square$

2.5 Ternaries with bounded multiplicities

2.5.1 Introduction

To support the guess that there exists only one counter-example we discuss Conjecture 2.1.1 for all ternary lattices with *bounded multiplicities*, i.e. there exists a number $b < \infty$ such that for all k the multiplicity satisfies

$$r(\gamma_k^2) := \# \left\{ x \in \mathbb{Z}^n : Q_\Gamma(x) = \gamma_k^2 \right\} \leq b.$$

Remark 2.5.1. While the multiplicities of a binary form are bounded (from above by 4) if and only if the form is non-arithmetic cf. [Kü1, p.166 Rem.], the ternary case is not that simple. Obviously a non-arithmetic ternary form that is arithmetic on a rational plane has unbounded multiplicities. Furthermore Kühnlein has shown in [Kü2, Cor.4.10] that a ternary form Q that does not admit an arithmetic plane has bounded multiplicities (from above by 8) if and only if the dimension of the $\mathbb{Q}$-subspace of $\mathbb{R}$ which is generated by the coefficients of Q, denoted by $\delta(Q)$, is at least 3.

First we discuss only orthogonal lattices with bounded multiplicities. Without doubt, orthogonality is a serious restriction, in particular the set of orthogonal lattices is a nullset in the set of lattices, and we do not really need this special case for the general one. But the following proof can guide us through the several cases of the general proof.

2.5.2 Orthogonal lattices

For this section, let Γ be an orthogonal ternary lattice with bounded multiplicities and covolume 1. Hence there exist $\mathbb{Q}$-linearly independent real numbers a, b, c with $a \cdot b \cdot c = 1$ and $0 < a < b < c$ such that

$$Q_\Gamma(x) = ax_1^2 + bx_2^2 + cx_3^2.$$

Lemma 2.5.2. *Let $b \leq 2$ and $c \leq 4$, then we have $\sigma_k > \gamma_k$ for all $k > 0$.*

Proof. Since $a < 1$ we have for all $x \in \mathbb{Z}^3$:

$$Q_\Gamma(x) = ax_1^2 + bx_2^2 + cx_3^2 \leq x_1^2 + 2x_2^2 + 4x_3^2 =: Q_\Psi(x).$$

In case that $Q_\Gamma(x) = Q_\Gamma(x')$ it follows that $x_i^2 = x_i'^2$ for $i = 1, 2, 3$ because a, b, c are linearly independent, so $Q_\Psi(x) = Q_\Psi(x')$. Thus the map

$$\gamma = Q_\Gamma(x) \longmapsto \psi = Q_\Psi(x)$$

is well-defined and surjective. Since $Q_\Gamma(x) \leq Q_\Psi(x)$ for all $x \in \mathbb{Z}^3$ we have

$$\gamma_k \leq \psi_k \text{ for all } k \geq 0.$$

Furthermore Q_Ψ represents a positive integer m if and only if $m \neq 4^a(16t + 14)$ for all $a, t \in \mathbb{N}_0$, cf. [Ram, p.13] resp. [Di1, Thm.V]. Accordingly Q_Ψ represents m if and only if $Q_{\tilde{\Sigma}}$ represents m:

$$\Rightarrow \psi_k = \tilde{\sigma}_k < \sqrt[6]{2} \cdot \tilde{\sigma}_k = \sigma_k \text{ for all } k > 0.$$

$\square$

Remark 2.5.3. Just like Q_Ψ the form $x_1^2 + x_2^2 + 2x_3^2$ represents an integer m if and only if $Q_{\tilde{\Sigma}}$ represents m. Hence we see here two more amusing examples for arithmetic lattices for which Conjecture 2.1.1 applies.

Proposition 2.5.4. *Let Γ be an orthogonal ternary lattice with bounded multiplicities and covolume 1. Then $\gamma_k < \sigma_k$ for all $k > 0$.*

Proof. Since $a < 1$ it follows that $a < \sigma_1^2$.
If $b < \sigma_2^2$ we have $a + b < \sigma_1^2 + \sigma_2^2 = \sigma_3^2$ and
if $b \geq \sigma_2^2$ we have $a = \frac{1}{bc} < \frac{1}{b^2} \leq \frac{1}{\sigma_2^4} < 0.158$, thus $a \cdot 4 < 0.632 < \sigma_2^2$ and $a \cdot 9 < 1.422 < \sigma_3^2$. Altogether:

$$\sigma_k > \gamma_k \text{ for } k = 1, 2, 3. \ (*)$$

Now consider the 2-dimensional sublattice

$$\Gamma' := \mathbb{Z} \begin{pmatrix} \sqrt{a} \\ 0 \end{pmatrix} \oplus \mathbb{Z} \begin{pmatrix} 0 \\ \sqrt{b} \end{pmatrix}.$$

Obviously $\gamma_k \leq \gamma'_k$. Kühnlein [Kü1, p.168] showed for the k-th length of an arbitrary reduced binary lattice with bounded multiplicities

$$L := \mathbb{Z} \begin{pmatrix} r \\ 0 \end{pmatrix} \oplus \mathbb{Z} \begin{pmatrix} s \\ t \end{pmatrix} :$$

$$l_k^2 \leq \frac{(4k + 2)\sqrt{\left(\frac{s}{r}\right)^2 + \left(\frac{t}{r}\right)^2}}{\pi\left(\frac{1}{r}\right)^2} = \frac{(4k + 2)r\sqrt{s^2 + t^2}}{\pi}.$$

For Γ' it follows:

$$\gamma'^2_k \leq \frac{4k + 2}{\pi}\sqrt{ab}.$$

The lower bound of σ_k of Proposition 2.3.2 dominates this upper bound for all $k \geq 4$ if $\sqrt{ab} \leq 0.742$ or $c \geq 1.817$ respectively. Hence together with $(*)$ the claim follows for $c \geq 1.817$.
If $c < 1.817$ we can use Lemma 2.5.2 because then $b < c < 2$. $\square$

2.5.3 General situation

Analogously to Proposition 2.5.4 we have:

Theorem 2.5.5. *Let Γ be a ternary lattice with bounded multiplicities and covolume 1. Then $\gamma_k < \sigma_k$ for all $k > 0$.*

The proof of this theorem will take the rest of Section 2.5.3. First we will avail ourselves again of Kühnlein's binary estimate to show in Lemma 2.5.6 and 2.5.9 that already the lengths of the binary sublattice

$$\Gamma' := \mathbb{Z}b_1 \oplus \mathbb{Z}b_2$$

are dominated by Σ if it is "small" enough. Apart from that if Γ' is "big" then the diameter of a fundamental parallelotope of Γ is bounded from above. Hence we are able to use volume arguments (Proposition 2.5.11) for these lattices to get an upper bound for their lengths. It remains to treat the first few lengths of these lattices with small diameter. Due to Lemma 2.5.8 it will suffice to consider the case that at least one of the coefficients a_{13}, a_{23}, a_{33} is independent from $Q_{\Gamma'}$. Then in every of these three cases we will just "count out" enough lengths that are not contained in the length spectrum of Γ'. Doubtless we will have to take great pains over this step (especially the last case), but after all it will close the proof.

Lemma 2.5.6. *Let $\Gamma := \mathbb{Z}b_1 \oplus \mathbb{Z}b_2 \oplus \mathbb{Z}b_3$ be a ternary lattice of covolume 1 given in a reduced form such that the binary sublattice $\Gamma' := \mathbb{Z}b_1 \oplus \mathbb{Z}b_2$ is non-arithmetic and $|b_1| \cdot |b_2| \leq 0.742$. Then $\gamma_k < \sigma_k$ for all $k > 0$.*

Proof. We take the rotated reduced lattice of the class of Γ such that

$$b_1 = \begin{pmatrix} r \\ 0 \\ 0 \end{pmatrix} \text{ and } b_2 = \begin{pmatrix} s \\ t \\ 0 \end{pmatrix}.$$

Hence we can use again Kühnlein's binary estimate as in the orthogonal case.
It remains to treat the first three lengths. Different from the proof of Proposition 2.5.4 we must be careful that we do not count a length twice. Now it is possible that the three basis vectors all have the same length, because then $\delta(Q)$ still can equal 3, cf. Remark 2.5.1.

Due to $|b_1| < \sqrt{0.742} < 0.862 < \sigma_1$ we have $2\,|b_1| < 1.723 < \sigma_3 = \sqrt[6]{2}\sqrt{3}$.

(a) If now $|b_1| < |b_2| < \sigma_2$ and $2\,|b_1| \neq |b_2|$ all is good. In case of $2\,|b_1| = |b_2|$ the length $|b_1 - b_2| \leq \sqrt{|b_1|^2 + |b_2|^2} \leq \sqrt{(\tfrac{1}{2}\sigma_2)^2 + \sigma_2^2} < 1.78 < \sigma_3$ and since Γ does not contain an arithmetic plane $|b_1 - b_2|$ is not commensurable with $|b_1|$ and $|b_2|$.

(b) If $|b_2| \geq \sigma_2$ then $|b_1| < \frac{0.742}{\sigma_2} < 0.468$, hence $2\,|b_1| < \sigma_2$ and $3\,|b_1| < \sigma_3$.

(c) If $|b_1| = |b_2|$ we pick out $|b_1 - b_2| \leq \sqrt{2|b_1|^2} < \sigma_2$ as the second length, since again a_{12} and $a_{11} = a_{22}$ are incommensurable.

$\square$

Corollary 2.5.7. *Let* $\Gamma := \mathbb{Z}b_1 \oplus \mathbb{Z}b_2 \oplus \mathbb{Z}b_3$ *be a ternary lattice of covolume 1 given in a reduced form such that the binary sublattice* Γ' *is non-arithmetic and* $|b_3| \geq \frac{\sqrt{2}}{0.742}$. *Then* $\gamma_k < \sigma_k$ *for all* $k > 0$.

Proof. With $e_3 \perp b_1, b_2$ we have

$$1 = \text{cov}(\Gamma) = \sin(\sphericalangle(b_1, b_2)) \cdot \cos(\sphericalangle(b_3, e_3)) \, |b_1| \, |b_2| \, |b_3| \, .$$

Denote $\omega := \sphericalangle(b_1, b_2)$. Since the basis is reduced (cf. Section 1.2) it is easy to see that

$$\sin(\omega) \cdot \cos(\sphericalangle(b_3, e_3)) \geq \sin(\omega) \cdot \sqrt{\frac{3}{4} - \frac{1 - \cos(\omega)}{4 \cdot (1 + \cos(\omega))}} \geq \frac{\sqrt{2}}{2} \, .$$

$\square$

Similarly in the case, that the multiplicities of the lengths of Γ' are bounded by 2, it follows that all lengths of Γ are dominated by those of Σ:

Lemma 2.5.8. *Let* $\Gamma := \mathbb{Z}b_1 \oplus \mathbb{Z}b_2 \oplus \mathbb{Z}b_3$ *be a ternary lattice of covolume 1 given in a reduced form such that the binary sublattice* Γ' *is non-arithmetic with* $\delta(Q_{\Gamma'}) = 3$. *Then* $\gamma_k < \sigma_k$ *for all* $k > 0$.

Proof. If $\delta(Q_{\Gamma'}) = 3$, then obviously the multiplicity of a nonzero length of Γ' is 2. Since the coefficients a_{11}, a_{12}, a_{22} are $\mathbb{Q}$-linearly independent two linearly independent vectors in Γ' cannot have the same length. Hence $2P_{\Gamma'}(X^2) - 1$ is at least the number of elements in Γ' of length $\leq X$. We will now give the binary estimate for this special case in more details (for the general calculation see [Kü1, p.167]). First we will estimate the number $2P_{\Gamma'}(X^2) - 1$ from above. To that end let x and y denote the coordinates of the lattice points with respect to the basis vectors $(1, 0)^T$ and $(\frac{s}{r}, \frac{t}{r})^T$. By solving the corresponding quadratic equation we have:

$$2P_{\Gamma'}(X^2) \geq \sum_{x=-[X]}^{[X]} \left(1 + \frac{2\sqrt{X^2((\frac{s}{r})^2 + (\frac{t}{r})^2) - x^2(\frac{t}{r})^2}}{(\frac{s}{r})^2 + (\frac{r}{t})^2} \right) + 1$$

$$\geq 2X + \frac{2}{\sqrt{(\frac{s}{r})^2 + (\frac{t}{r})^2}} \sum_{x=-[X]}^{[X]} \sqrt{X^2 - x^2}$$

$$\geq 2X + \frac{2}{\sqrt{(\frac{s}{r})^2 + (\frac{t}{r})^2}} \left[\int_{-X}^{X} \sqrt{X^2 - x^2} \, dx - X \right]$$

$$\geq \frac{X^2 \pi}{\sqrt{(\frac{s}{r})^2 + (\frac{t}{r})^2}} \, .$$

This gives

$$\tilde{\gamma}' \left[\frac{X^2\pi}{2 \cdot \sqrt{(\frac{s}{r})^2 + (\frac{t}{r})^2}} \right] \leq X$$

and thus for the normalised k-th length

$$(\gamma'_k)^2 \leq \frac{2k+2}{\pi} r\sqrt{s^2 + t^2}.$$

Due to the proof of Corollary 2.5.7 and Lemma 2.5.6 we can assume that

$$|b_1| \cdot |b_2| \leq \frac{\sqrt{2}}{|b_3|} \leq \frac{\sqrt{2}}{|b_2|} \leq \frac{\sqrt{2}}{\sqrt{0.742}} < 1.64.$$

Hence we have

$$(\gamma'_k)^2 \leq \frac{2k+2}{\pi} \cdot 1.64.$$

The lower bound of σ_k dominates this upper bound for all $k \geq 8$. Compared to the exact values of Σ we even have a dominance for $k \geq 5$. The first four lengths are quite simple in this case because two linearly independent elements of Γ' have different lengths:
We always have $|b_1| < \sigma_1$, if $|b_2| < \sigma_2$ then $|b_1 - b_2| < \sigma_3$ and $2|b_1| < \sigma_4$. If $|b_2| \geq \sigma_2$ we have again due to the proof of Corollary 2.5.7 $|b_1| \leq \frac{\sqrt{2}}{|b_2| \cdot |b_3|} \leq \frac{\sqrt{2}}{|b_2|^2} \leq \frac{\sqrt{2}}{\sigma_2^2} < 0.561$. Hence $2|b_1| < \sigma_2$, $3|b_1| < \sigma_3$ and $|b_2| \leq |b_3| < 1.906 < \sigma_3 < \sigma_4$. □

In contrast to that the general binary estimate becomes ineffective when $|b_1| \cdot |b_2|$ grows. Strictly spoken if $|b_1| \cdot |b_2| > \sqrt[3]{2} \cdot \frac{17}{16} \cdot \frac{\pi}{4} \approx 1.0514$ it always dominates the upper bound of Σ asymptotically. Unfortunately we are not able to estimate the lengths using a comparison to a nice arithmetic form as in the orthogonal case. But as the first length grows the complete length spectrum of Γ draws near to the complete length spectrum of Σ, hence we can use volume arguments cf. Section 3.1. To simplify these matters we will go a further step with the binary estimate in the next lemma.

Lemma 2.5.9. *Let $\Gamma := \mathbb{Z}b_1 \oplus \mathbb{Z}b_2 \oplus \mathbb{Z}b_3$ be a ternary lattice with bounded multiplicities and covolume 1 given in a reduced form such that $|b_1| \cdot |b_2| \leq 1$. Then $\gamma_k < \sigma_k$ for all $k > 0$.*

Proof. Due to Lemma 2.5.6 and Lemma 2.5.8 we are able to assume for the proof that $|b_1| \cdot |b_2| > 0.742$ and that $\delta(Q_{\Gamma'}) = 2$. Furthermore we claim that one of the lengths $|b_3|$, $|b_1 - b_3|$, $|b_2 - b_3|$ is not an element of the length spectrum of Γ'. If $|b_3|$ is a length of Γ' then the dimension of the $\mathbb{Q}$-subspace which is generated by $a_{11}, a_{12}, a_{22}, a_{33}$, denoted by $\delta(a_{11}, a_{12}, a_{22}, a_{33})$, equals $\delta(Q_{\Gamma'}) = 2$. Since $\delta(Q_\Gamma) \geq 3$ it follows that a_{13} or a_{23} cannot be written as a linear combination of a_{11}, a_{12}, a_{22}. Hence $|z_1 b_1 - z_3 b_3|$ or $|z_2 b_2 - z_3 b_3|$ is no length of Γ' for all $z_1, z_2, z_3 \in \mathbb{Z} \setminus \{0\}$. (Without loss of generality we assume $a_{23} \geq 0$, otherwise replace b_3 by $-b_3$.)

According to that if for a k_0 the length $|b_3|$ or $|b_1 - b_3|$ or $|b_2 - b_3|$ is less than σ_{k_0}, then we have for $k \geq k_0$

$$\gamma_k^2 \leq (\gamma'_{k-1})^2 \leq \frac{4k-2}{\pi} |b_1| \cdot |b_2| \leq \frac{4k-2}{\pi}.$$

This new upper bound is dominated by the bound of Σ for $k \geq 8$. If we compare the bound with the first seven exact values of Σ, cf. Table 2.4, we can see that it remains to deal with the lengths less than σ_{k_0}.

Table 2.4: Lengths 2-7

k	2	3	4	5	6	7
$\frac{4k-2}{\pi}$	1.91	3.18	4.46	5.73	7.01	8.28
σ_k^2	2.52	3.78	5.03	6.30	7.56	8.82

In any case we can again assume that

$|b_3| < \sigma_3$, $|b_1 - b_3| \leq \sqrt{|b_1|^2 + |b_3|^2} \leq \sqrt{\sqrt[3]{2} + 1.906^2} < 2.22 < \sigma_4$ and $|b_2 - b_3| \leq \sqrt{2} \cdot 1.906 < 2.70 < \sigma_6$. To be on the safe side we therefore have to collect five lengths, again we must keep in mind not to count a length twice:

(a) Let now $|b_1| < |b_2| < \sigma_2$ then $|b_2 - b_3| < \sqrt{\sigma_2^2 + 1.906^2} < 2.48 < \sigma_5$.
   If now $|b_2| \neq |b_3|$ we can take $|b_3|$ for the third length and $2|b_1|$ for the fourth in case that $2|b_1|$ does not equal $|b_2|$ and $|b_3|$. Otherwise we can choose $|b_1 - b_2|$ or $|b_1 - b_3| < \sqrt{1 + 1.906^2} < \sigma_4$ because Γ does not contain an arithmetic plane.
   If on the other side $|b_2| = |b_3|$, and so we have already counted this length, then $|b_1 - b_3| \leq \sqrt{1 + \sigma_2^2}$ is less than σ_3 and if $|b_2| < 1.374$ then also $|b_2 - b_3| < \sqrt{2|b_2|^2} < 1.9432 < \sigma_3$.
   And in case of $|b_2| \geq 1.374$ we have $|b_1| < \frac{1}{1.374}$ and so we find either $2|b_1|$ or $|b_1 - b_2|$ as the third length. Finally in this case $|b_2 - b_3| < \sqrt{2\sigma_2^2} = \sigma_4$.

(b) If $|b_2| \geq \sigma_2$ then again $|b_1| < 0.561$ hence $2|b_1| < \sigma_2$, $3|b_1| < \sigma_3$ and $4|b_1| < \sigma_5$.
   If now $|b_2|$ and $|b_1|$ are incommensurable then $|b_2|$ can be choosen for the fourth length.
   If $|b_1|$ and $|b_2|$ are commensurable then $|b_1 - b_2|$ does not equal multiples of $|b_1|$ and is less than σ_4. (This holds analogously for $|b_3|$.)

(c) If finally $|b_1| = |b_2|$, both lengths $|b_1 - b_3|$ and $|b_2 - b_3|$ are less than σ_4. For the second length we pick $|b_1 - b_2| < \sqrt{2\sigma_1^2}$ and for the third length $|b_3|$,
   or in case that we have already previously counted $|b_3|$ with $|b_1|$ or $|b_1 - b_2|$ both lengths $|b_1 - b_3|$, $|b_2 - b_3|$ are already less than σ_3.

$\square$

Corollary 2.5.10. *Let $\Gamma := \mathbb{Z}b_1 \oplus \mathbb{Z}b_2 \oplus \mathbb{Z}b_3$ be a ternary lattice with bounded multiplicities and covolume 1 given in a reduced form such that $|b_3| \geq \sqrt{2}$. Then $\gamma_k < \sigma_k$ for all $k > 0$.*

Proposition 2.5.11. *Let Γ be a ternary lattice with bounded multiplicities and denote by d the diameter of a fundamental parallelotope and by $B(X)$ the 3-dimensional ball with radius $\sqrt{X}$. Then we have*

$$P_\Gamma(X) \geq \frac{1}{8}(\mathrm{vol}(B((\sqrt{X} - \frac{d}{2})^2)) + 7).$$

Proof. Since the multiplicities of Γ are bounded by 8 and due to the fact that the multiplicity of 0 is 1, $8P_\Gamma(X) - 7$ is at least the number of elements in Γ of length less than $\sqrt{X}$. And this number is greater than $\mathrm{vol}(B((\sqrt{X} - \frac{d}{2})^2))$, cf. proof of Theorem 3.1.1. □

If for some $X_0 > 0$ the inequality $P_\Gamma(X) > P_\Sigma(X)$ holds for all $X \geq X_0$, then

$$\sigma_{P_\Sigma(X)} > \gamma_{P_\Gamma(X)} > \gamma_{P_\Sigma(X)} \text{ for all } X \geq X_0.$$

Hence $\lambda_k < \sigma_k$ for all $k \geq P_\Sigma(X_0)$.
With the bound of Proposition 2.3.6 we get

$$
\begin{aligned}
P_\Gamma(X) - P_\Sigma(X) &\geq \tfrac{\pi}{6}\left(X^{\frac{3}{2}} - 3\tfrac{d}{2}X + 3\tfrac{d^2}{4}X^{\frac{1}{2}} - \tfrac{d^3}{8} + \tfrac{21}{4\pi}\right) - \left(\tfrac{1}{\sqrt[3]{2}}\tfrac{15}{16}X + \tfrac{14}{16}\right) \\
&= \tfrac{\pi}{6}X^{\frac{3}{2}} - \left(\tfrac{1}{\sqrt[3]{2}}\tfrac{15}{16} + \tfrac{\pi}{2}\cdot\tfrac{d}{2}\right)X + \tfrac{\pi}{2}\cdot\tfrac{d^2}{4}X^{\frac{1}{2}} - \tfrac{\pi}{6}\cdot\tfrac{d^3}{8}.
\end{aligned}
$$

We take note that $P_\Gamma(15.57(\frac{d}{2})^2) - P_\Sigma(15.57(\frac{d}{2})^2) > 0$ because $d \geq \sqrt{3}$, the minimal diameter of the cubic lattice.

Now let Γ be such that $|b_1| \cdot |b_2| > 1$, cf. Lemma 2.5.9. Hence we get
$|b_1| \cdot |b_2| = (\sin(\sphericalangle(b_1, b_2)) \cdot \cos(\sphericalangle(b_3, e_3)) \cdot |b_3|)^{-1} \leq \sqrt{2} \cdot \frac{1}{|b_2|} \leq \sqrt{2}$ and also $|b_1| \cdot |b_3| \leq \sqrt{2}$, so

$$
\begin{aligned}
d^2 &= |b_1|^2 + |b_2|^2 + |b_3|^2 + 2|b_1| \cdot |b_2| \cdot \cos(\sphericalangle(b_1, b_2)) \\
&\quad + 2|b_1| \cdot |b_3| \cdot \cos(\sphericalangle(b_1, b_3)) + 2|b_2| \cdot |b_3| \cdot \cos(\sphericalangle(b_2, b_3)) \\
&\leq |b_1|^2 + |b_1||b_2| + |b_1||b_3| + 3|b_3|^2 \leq \sqrt[3]{2} + 2\sqrt{2} + 6.
\end{aligned}
$$

Hence $\lambda_k < \sigma_k$ for all $k \geq P_\Sigma(15.57 \cdot \frac{\sqrt[3]{2}+2\sqrt{2}+6}{4})$, and this number is less than 31. It remains (the ugly part of the proof) to collect the first 30 lengths.

Therefore we play the same game as in the proof of Lemma 2.5.9, but repeatedly. Again we keep in mind that we can assume that $|b_1| \cdot |b_2| > 1$ and that (due to Lemma 2.5.8) we are able to exclude that $\delta(Q_{\Gamma'}) = 3$.

$(\mathbf{A})$ First we consider the case that $\delta(a_{11}, a_{12}, a_{22}, a_{33}) = 2$ and a_{13} cannot be written as Q-linear combination of the coefficients of $Q_{\Gamma'}$. As already discussed above then for all $z_1, z_3 \in \mathbb{Z} \smallsetminus \{0\}$ we have $|z_1 b_1 + z_3 b_3| \neq |\gamma'|$ for all $\gamma' \in \Gamma'$. And for two such lengths $|z_1 b_1 + z_3 b_3|$ and $|z_1' b_1 + z_3' b_3|$ to be equal we necessarily have $z_1 z_3 = z_1' z_3'$. Therefore we avoid problems if we always pick out only one length with the given product $z_1 z_3$. Table 2.5 gives enough lengths to conclude this case, whereas we use again that Γ is reduced, for example by the estimate $|b_1 + b_3|^2 \leq |b_1|^2 + |b_3|^2 + 2 |b_1| |b_3| \cos(\sphericalangle(b_1, b_3)) < \sqrt[3]{2} + 2 + \sqrt{2} < \sigma_4$.

Table 2.5: Lengths $\notin \Gamma'$

$z_1 z_3$	-1	1	-2	2
	$|b_1 - b_3| < \sigma_3$	$|b_1 + b_3| < \sigma_4$	$|2b_1 - b_3| < \sigma_6$	$|2b_1 + b_3| < \sigma_8$
	-3	-4	3	4
	$|3b_1 - b_3| < \sigma_{11}$	$|2b_1 - 2b_3| < \sigma_{11}$	$|3b_1 + b_3| < \sigma_{14}$	$|2b_1 + 2b_3| < \sigma_{14}$
	-6	-5	6	5
	$|3b_1 - 2b_3| < \sigma_{15}$	$|5b_1 - b_3| < \sigma_{26}$	$|3b_1 + 2b_3| < \sigma_{26}$	$|5b_1 + b_3| < \sigma_{31}$

Again $|b_1| < \sigma_1$, $|b_2|$ or $|b_1 - b_2| < \sigma_2$, and similarly as in previous cases we can choose either $2|b_1|$ or $3|b_1|$ or $|2b_1 - b_2|$ for the fifth length. With $|b_1 - b_3|$, $|b_1 + b_3|$ and $|2b_1 - b_3|$ we have three different lengths that are not in the length spectrum of Γ' and that are less than σ_3, σ_4 and σ_6 respectively. Hence for $k \geq 7$ we have

$$\gamma_k^2 \leq (\gamma_{k-3}')^2 \leq \frac{4k - 10}{\pi} |b_1| |b_2| \leq \frac{4k - 10}{\pi} \sqrt{2}.$$

If we compare this new upper bound with the exact values σ_k we have a dominance for $k \leq 8$. Continuing, the length $|2b_1 + b_3|$ is less than σ_8, hence for $k \geq 9$

$$\gamma_k^2 \leq (\gamma_{k-4}')^2 \leq \frac{4k - 14}{\pi} \sqrt{2}.$$

This new upper bound is dominated by the bound of Σ for $k \leq 11$. $|3b_1 - b_3|$ and $|2b_1 - 2b_3| < \sigma_{11}$, hence for $k \geq 12$

$$\gamma_k^2 \leq (\gamma_{k-6}')^2 \leq \frac{4k - 22}{\pi} \sqrt{2}.$$

This new upper bound is dominated by the bound of Σ for $k \leq 19$. The lengths $|3b_1 + b_3|$, $|2b_1 + 2b_3|$ and $|3b_1 - 2b_3|$ are less than σ_{19}, hence for $k \geq 20$

$$\gamma_k^2 \leq (\gamma_{k-9}')^2 \leq \frac{4k - 34}{\pi} \sqrt{2}.$$

This upper bound finally is dominated by the bound of Σ for $k \leq 30$, which concludes this case.

(B) Next we discuss the case that all lengths $|z_2b_2 - z_3b_3|$ are not in the length spectrum of Γ', to be exact that $\delta(a_{11}, a_{12}, a_{22}, a_{33}) = 2$ and $\delta(a_{11}, a_{12}, a_{22}, a_{23}) = 3$. Since we can only assume that $|b_2| \leq |b_3| < \sqrt{2}$ the estimates of these lengths are always greater. Hence we have to collect more carefully, i.e. we need additional lengths that are not elements of the length spectrum of Γ'. Due to case (A) we can also assume that $\delta(a_{11}, a_{12}, a_{22}, a_{13}) = 2$. Hence all lengths $|z_1b_1 + z_2b_2 + z_3b_3|$ with $z_1, z_2, z_3 \in \mathbb{Z} \setminus \{0\}$ are not contained in the length spectrum of Γ'. And such a length only equals $z_2'b_2 + z_3'b_3$ if $z_2z_3 = z_2'z_3'$. For $z_2z_3 = 1$ we pick $|b_1 + b_2 + b_3| < \sqrt{\sqrt[3]{2} + 2\sqrt{2} + 6} < \sigma_9$, but $|b_2 + b_3|$ is the only length used with $z_2z_3 = 1$. If they were equal it would follow that $a_{11} + 2a_{12} + 2a_{13} = 0$ and that contradicts $a_{11} > 0$ and $a_{12}, a_{13} \geq 0$. The second additional length we need is $|b_1 + b_2 - b_3| < \sqrt{\sqrt[3]{2} + 4 + \sqrt{2}} < \sigma_6$. If it equals $|b_2 - b_3|$ then $a_{11} + 2a_{12} - 2a_{13} = 0$. Since additionally $2a_{13} \leq a_{11}$ it follows that $2a_{13} = a_{11}$ and $a_{12} = 0$. If that is the case we pick $|b_1 - b_2 + b_3| < \sigma_6$, where the conditions would be $2a_{12} = a_{11}$ and $a_{13} = 0$.

Furthermore we have the lengths $|b_2 - b_3| < \sigma_4$, $|b_2 + b_3| < \sigma_5$ and $|2b_2 - b_3| < \sigma_8$. Hence to start the game again we must treat the first three and the seventh length. As always $|b_1| < \sigma_1$, and if $|b_1| = |b_2|$ then the estimates are the same as in the first case. Thus let $|b_1| \neq |b_2|$, hence our second length is $|b_2|$. If now $|b_2| \neq |b_3|$ this is the third and $2|b_3| < \sigma_7$. If $|b_2| = |b_3|$ then $|b_1 - b_2| \neq |b_1|, |b_2|$ is less than σ_3 and again we can surely choose $2|b_3| < \sigma_7$.

Hence we get for $k \geq 10$

$$\gamma_k^2 \leq (\gamma_{k-5}')^2 \leq \frac{4k - 18}{\pi} |b_1| |b_2| \leq \frac{4k - 18}{\pi} \sqrt{2}.$$

This new upper bound is dominated by the bound of Σ for $k \leq 15$. Continuing, the lengths $|2b_2 + b_3|$, $|2b_2 - 2b_3|$ and $|3b_2 - b_3|$ are less than σ_{15}. Hence for $k \geq 16$

$$\gamma_k^2 \leq (\gamma_{k-8}')^2 \leq \frac{4k - 30}{\pi} \sqrt{2}.$$

This upper bound now is dominated by the bound of Σ for $k \leq 26$. Finally the length $|2b_2 + 2b_3|$ is less than σ_{26}. Hence for $k \geq 27$ the k-th length of Γ is less than γ_{k-9}'.

(C) The last case is that already $\delta(a_{11}, a_{12}, a_{22}, a_{33}) = 3$. Here it is easy to find plenty of lengths that are not lengths of Γ', but it is not so obvious that they are all different. Therefore we are not able to avoid some more cases.

At first $z \cdot |b_3|$ does not lie in the length spectrum of Γ' for all $z \in \mathbb{N}$. For our game this always gives us three lengths, less than σ_2, σ_7 and σ_{14}.

Furthermore we remember Lemma 2.5.8: since $|b_1| \cdot |b_3| \leq \sqrt{2} < 1.64$ there is nothing to show if $\delta(a_{11}, a_{13}, a_{33}) = 3$. Thus let $\delta(a_{11}, a_{13}, a_{33}) = 2$. As assumed in this case a_{11} and a_{33} are incommensurable, hence a_{13} can be uniquely written as

$$a_{13} = q_1 a_{11} + q_3 a_{33}, \quad q_1, q_3 \in \mathbb{Q}.$$

Therefore we have

$$|z_1 b_1 + z_3 b_3|^2 = (z_1^2 + 2z_1 z_3 q_1) a_{11} + (z_3^2 + 2z_1 z_3 q_3) a_{33}.$$

In case of $q_1, q_3 \neq \frac{1}{2}$ the lengths $z \cdot |b_1 - b_3|$, $z \in \mathbb{N}$, are incommensurable with a_{11} and with a_{33}, hence they are not lengths of Γ' and do not equal $z' \cdot |b_3|$. This gives three more lengths, less than σ_3, σ_{11} and σ_{23} respectively.

Next we look at the sublattice $\mathbb{Z}b_2 \oplus \mathbb{Z}b_3$. If $\delta(a_{22}, a_{23}, a_{33}) = 3$ we can employ all lengths $|z_2 b_2 + z_3 b_3|$, the interesting case is that $a_{23} = q'_2 a_{22} + q'_3 a_{33}$. If $q'_2, q'_3 \neq \frac{1}{2}$ then analogously we can choose the lengths $z \cdot |b_2 - b_3|$, less than σ_4 and σ_{13}. To avoid still more cases we consider now $q'_2 = \frac{1}{2}$, then $|z_2 b_2 + z_3 b_3|^2 = (z_2^2 + z_2 z_3) a_{11} + (z_3^2 + 2z_1 z_3 q_3) a_{33}$. Hence we can choose $(z_2, z_3) = (2, -1), (1, -2)$ if $q'_{33} \leq 0$ and $(z_2, z_3) = (2, 1), (1, 2)$ if $q'_{33} > 0$ respectively. Then the coefficients of $(z_2^2 + z_2 z_3)$ and $(z_3^2 + 2z_1 z_3 q_3)$ surely do not equal zero and are all different. These lengths are less than σ_{12}. Thus we do not make a mistake with this sublattice if we count two lengths less than σ_{12} and σ_{13} (obviously we can change the roles of q'_2 and q'_3).

We need at least one more (small) length. If $q_1, q_3 \neq -\frac{1}{2}$ then also $|b_1 + b_3|$ is not a length of Γ' and is unequal to $z \cdot |b_3|$, but it could equal one of the already counted lengths $z \cdot |b_1 - b_3|$. On the one hand $|b_1 + b_3| < 2.16$ and on the other hand $z \cdot |b_1 - b_3| > z\sqrt{\frac{3}{2}}$, thus $z = 1$ remains the only possibility, hence we exclude now the case $q_1 = q_3 = 0$. Then $|b_1 + b_3|$ is the next length less than σ_4.
In the excluded case of $q_1 = q_3 = 0$ we can see by pluging in that $|2b_1 - b_3|$ is not used already and do not lie in the length spectrum of Γ' (remember that a_{11} and a_{33} are incommensurable). Furthermore if $q_1 = -\frac{1}{2}$ (or $q_3 = -\frac{1}{2}$) then $q_3 > 0$, otherwise a_{13} would be negative. Hence $|2b_1 + b_3|^2 = 2a_{11} + (1 + 4q_3) a_{33}$ fulfils all conditions since $1 + 4q_3 \neq 0$ and no used length has a_{11}-part 2. Both of these lengths are less than σ_8, thus in all cases we are able to count one last length which certainly is less than σ_8.

Now we have collected 9 lengths and can start the game:

We simplify matters for the first lengths if we observe that the cases (A) and (B) also hold for the sublattice $\mathbb{Z}b_1 \oplus \mathbb{Z}b_3$, hence we can assume that a_{11} is incommensurable with a_{22} too. Then $|b_1| < \sigma_1$, $|b_2| < \sigma_4$ and $2|b_1| < \sigma_5$. With analogous arguments as before one of the lengths $|b_1 - b_2|$, $|b_1 + b_2|$, $|2b_1 + b_2|$ is independent from $|b_1|$ and $|b_2|$ and all of them are less than σ_6. Additionally we have found lengths less than $\sigma_2, \sigma_3, \sigma_7$ and σ_8 that are not in the length spectrum of Γ'. Hence we get a new bound which is dominated for $k \leq 11$. Next we have lengths less than σ_{11}, σ_{12}, σ_{13} and σ_{14}. Thus we get a new bound for $k \geq 15$ with $\gamma_k \leq \gamma'_{k-8}$. This gives a dominance for $k \leq 26$ and finally the 9-th length is less than σ_{23}.

It remains to deal with the case q_1 or q_3 equals $\frac{1}{2}$. We still can use the lengths $z \cdot |b_3|$ and the lengths from $\mathbb{Z}b_2 \oplus \mathbb{Z}b_3$. We now start with the first lengths: $|b_1| < \sigma_1$, $|b_2| < |b_3| < \sigma_2$ and $2|b_1| < \sigma_4$ (obviously they are different). Furthermore the length $|b_1 + b_2|$ either equals $2|b_2|$ or it is incommensurable with $|b_1|$ and with $|b_2|$, hence we have the fifth length. Similarly we can use $|b_1 - b_3|$ or $|b_1 + b_3|$ or in case that $(q_1, q_3) = (\frac{1}{2}, -\frac{1}{2})$ we choose $|2b_1 + b_3|^2 = 6a_{11} - a_{33}$ as the sixth length, which is not a length of Γ'. The rest follows as usual, let $q_1 = \frac{1}{2}$ then

$$|z_1 b_1 + z_3 b_3|^2 = (z_1^2 + z_1 z_3)a_{11} + (z_3^2 + 2z_1 z_3 q_3)a_{33}.$$

Hence we choose $(z_1, z_3) = (2, 1), (1, 2), (3, 1)$, or $(z_1, -z_3)$ if $q_3 < 0$. Since a_{11} and a_{33} are incommensurable it is easy to be seen that these lengths are all different and are either no lengths of Γ' and that they are not already used. Obviously this holds analogously for $q_3 = \frac{1}{2}$. Hence we have four additional lengths that are less than σ_6, σ_8, σ_{10} and σ_{14}. Finally in this case we have collected 9 lengths that run through the bounds as before.

All together we have finished the proof of Theorem 2.5.5.

Remark 2.5.12. Together with Proposition 2.3.5 we get the amusing fact that for all ternary lattices Γ with bounded multiplicities the sixth length γ_6 is less than the sixth length λ_6 of the honeycomb lattice.

2.5.4 Higher dimensions

In contrast to the binary and ternary case we do not know exactly how the lattices with bounded multiplicities in higher dimensions look like. So the ideas of the ternary proof will not give us a good approach, but a criterion, when a lattice of a higher dimension with bounded multiplicities certainly fulfils Conjecture 2.1.1.

For effective results we first need an upper bound for the lengths of Σ_3. Here as before Σ_n denotes the normalised lattice with the best known sphere packing in dimension $2 \leq n \leq 8$ cf. Table 2.1 and, if it is clear in which dimension we are, σ_k its k-th length.

Proposition 2.5.13. *For the k-th length σ_k of Σ_3 we have:* $\sigma_k \leq \alpha \cdot \sqrt{\frac{600}{547}k}$.

Proof. Following the proof of Proposition 2.3.4 we have
$\#\{t \in \mathbb{N}_0 : 4^a(16t + 14) \le k\} = 0$ if $4^a \cdot 14 > k$ for $a \in \mathbb{N}_0$, hence

$$\# \{(a,t) \in \mathbb{N}_0^2 : 4^a(16t+14) \le k\}$$

$$= \sum_{a=0}^{\left[\log_4\left(\frac{k}{14}\right)\right]} \# \{t \in \mathbb{N}_0 : 4^a(16t+14) \le k\}$$

$$= \sum_{a=0}^{\left[\log_4\left(\frac{k}{14}\right)\right]} \left[\tfrac{k-4^a\cdot 14}{4^a\cdot 16} + 1\right] \le \sum_{a=0}^{\left[\log_4\left(\frac{k}{14}\right)\right]} \left(\tfrac{k}{4^a\cdot 16} + \tfrac{1}{8}\right)$$

$$\le \tfrac{k}{16}\cdot \underbrace{\sum_{a=0}^{\infty} \tfrac{1}{4^a}}_{=\frac{1}{1-\frac{1}{4}}} + \sum_{a=0}^{\left[\log_4\left(\frac{k}{14}\right)\right]} \tfrac{1}{8} = \tfrac{1}{12}k + \tfrac{1}{8\ln(4)}\ln(k) - \tfrac{\ln(14)}{8\ln(4)}\;.$$

It is easy to see that $\frac{1}{\ln(4)}\ln(k) - \frac{\ln(14)}{\ln(4)} \le \frac{1}{25}k$ for all $k \in \mathbb{N}$ and this estimate will do all that we need:

$$\# \{(a,t) \in \mathbb{N}_0^2 : 4^a(16t+14) \le k\} \le \tfrac{1}{12}k + \tfrac{1}{8}\cdot\tfrac{1}{25} = \tfrac{53}{600}k \; (*)$$

$$\Rightarrow \tilde{\lambda}_k^2 \le k + \#\{(a,t) \in \mathbb{N}_0^2 : 4^a(16t+14)\} \le k + \tfrac{53}{600}k\sum_{j=0}^{\infty}\left(\tfrac{53}{600}\right)^j\}$$

$$\le k + \#\{(a,t) \in \mathbb{N}_0^2 : 4^a(16t+14)\} \underset{(*)}{\le} 1\tfrac{53}{547}k\} \le 1\tfrac{53}{547}k.$$

$\square$

Corollary 2.5.14. *Let Γ be a quaternary lattice with covolume 1 such that there exists a ternary sublattice Γ' of Γ with bounded multiplicities and $\mathrm{cov}(\Gamma')^2 \le \sqrt{2}(\frac{547}{600})^3 \approx 1.0716$. Then the lengths of Σ_4 dominate the lengths of Γ.*

Proof. The form $Q_{\widetilde{\Sigma}_4}$ is universal and its determinant is $\frac{1}{4}$, hence $\sigma_k^2 = \sqrt{2}k$. With Theorem 2.5.5 and Proposition 2.5.13 follows

$$\gamma_k^2 \le (\gamma_k')^2 < \sqrt[3]{\mathrm{cov}(\Gamma')^2} \cdot \alpha^2 \cdot \frac{600}{547}k \le \sqrt[6]{2}\cdot\frac{547}{600}\cdot\sqrt[3]{2}\cdot\frac{600}{547}k = \sqrt{2}k.$$

$\square$

Obviously we can give this criterion in a more general way:

Corollary 2.5.15. *Let $\widetilde{\Xi}$ be an integral universal n-dimensional lattice, $n \ge 4$, Γ an n-dimensional lattice of covolume 1 with a ternary sublattice Γ' with bounded multiplicities such that*

$$\mathrm{cov}(\Gamma')^2 \le \left(\frac{1}{\alpha^2}\cdot\frac{547}{600}\cdot\mathrm{cov}(\widetilde{\Xi})^{-\frac{2}{n}}\right)^3.$$

Then the lengths of the normalised lattice Ξ dominate the lengths of Γ.

Proof. Similar to Corollary 2.5.14 we have

$$\gamma_k^2 \le (\gamma_k')^2 < \sqrt[3]{\operatorname{cov}(\Gamma')^2} \cdot \alpha^2 \cdot \frac{600}{547}k \le \operatorname{cov}(\widetilde{\Xi})^{-\frac{2}{n}} \cdot k = \xi_k.$$

$\square$

With regard to Conjecture 2.1.1 the bounds of the covolume of Γ' for Σ_n $(4 \le n \le 8)$ are given in Table 2.6.

Table 2.6: Bounds of $\operatorname{cov}(\Gamma')$

n	4	5	6	7	8
$\operatorname{cov}(\widetilde{\Sigma}_n)^2$	$\frac{1}{4}$	$\frac{1}{8}$	$\frac{3}{64}$	$\frac{1}{64}$	$\frac{1}{256}$
$\operatorname{cov}(\Gamma')^2 \le$	1.0715	1.3192	1.7498	2.2519	3.0308

Chapter 3

Complete length spectrum

3.1 Introduction

It may be surprising that in Chapter 2 the multiplicities of the lengths do not play a role at all. This is contrary to the usual definition of the length spectrum of a surface. But Schmutz Schaller showed in [Schm1, p.202] that in the hyperbolic case the existence of a surface with maximal (complete) length spectrum is excluded. More precisely he showed that for all surfaces M of genus $g \geq 2$ there exists a surface M' in a small neighbourhood of M in the moduli space of surfaces of genus g, such that at least one of the closed geodesics is longer in M' than in M.

In this chapter we discuss the corresponding Euclidean results. Therefore once again we sort the lengths of the elements of a lattice Γ in $\mathbb{R}^n$ according to size, but now considering the multiplicities of these lengths:

$$0 = \gamma_0^{(c)} < \gamma_1^{(c)} \leq \gamma_2^{(c)} \leq \gamma_3^{(c)} \leq \gamma_4^{(c)} \leq \cdots$$

where in the sequel $\gamma_k^{(c)}$ denotes the *k-th complete length* of Γ.

Since for all lattices with the same covolume the number of lattice points whose lengths do not exceed X behave asymptotically similar, we cannot even expect an asymptotical dominance as in Section 2.2. Crude volume arguments give:

Theorem 3.1.1. *Let Γ be a lattice with basis $\{b_1, ..., b_n\}$ and $B_n(X)$ the n-dimensional ball with radius $\sqrt{X}$. Then we have*

$$(a) \quad S_\Gamma(X) \;:=\; \#(\Gamma \cap B_n(X)) = \tfrac{1}{\mathrm{cov}(\Gamma)}\mathrm{vol}(B_n(X)) + \mathcal{O}(X^{\frac{n-1}{2}})$$

$$(b) \quad S_\Gamma(X) \;=\; \tfrac{1}{\mathrm{cov}(\Gamma)}\mathrm{vol}(B_n(X)) + \Omega(X^{\frac{n}{2}-1}), \;\; \textit{if } \Gamma \textit{ is arithmetic.}$$

Proof. At this place we give a brief proof, for further information see [Fr] and the historical remarks given therein.

(a) We consider for every lattice point $\gamma = \sum\limits_{i=1}^{n} \gamma_i b_i \in \Gamma$ the parallelotope

$$\vartheta(\gamma) := \{x \in \mathbb{R}^n : \forall i = 1, ..., n : \gamma_i - \frac{1}{2}|b_i| \leq x_i \leq \gamma_i + \frac{1}{2}|b_i|\}$$

and build the polyhedron

$$P(X) := \bigcup_{\gamma \in B_n(X)} \vartheta(\gamma)$$

with $\mathrm{vol}(P(X)) = \mathrm{cov}(\Gamma)S_\Gamma(X)$. For this polyhedron we have the inclusions (assume $\sqrt{X} > \frac{d}{2}$)

$$B_n((\sqrt{X} - \frac{d}{2})^2) \subset P(X) \subset B_n((\sqrt{X} + \frac{d}{2})^2),$$

where d again is the diameter of a fundamental parallelotope. Now, if we compare the volumes we obtain

$$\mathrm{vol}(B_n((\sqrt{X} - \frac{d}{2})^2))\ \leq \mathrm{vol}(P(X)) \qquad \leq \mathrm{vol}(B_n((\sqrt{X} + \frac{d}{2})^2)),$$

$$\frac{\pi^{\frac{n}{2}}}{\Gamma(\frac{n}{2}+1)}(\sqrt{X} - \frac{d}{2})^n \qquad \leq \mathrm{cov}(\Gamma)S_\Gamma(X)\ \leq \frac{\pi^{\frac{n}{2}}}{\Gamma(\frac{n}{2}+1)}(\sqrt{X} + \frac{d}{2})^n,$$

here Γ is the *Gamma function*.

(b) Assume for an integral lattice Γ (for simplicity let $\mathrm{cov}(\Gamma) = 1$)

$$\lim_{X \to \infty} \frac{\mathrm{cov}(\Gamma)S_\Gamma(X) - \mathrm{vol}(B_n(X))}{X^{\frac{n}{2}-1}} = 0.$$

Since $S_\Gamma(X) = S_\Gamma([X])$ we have for $N \in \mathbb{N}$

$$\frac{\mathrm{vol}(B_n(N + \frac{1}{2})) - \mathrm{vol}(B_n(N))}{N^{\frac{n}{2}-1}}$$

$$= \frac{S_\Gamma(N) - \mathrm{vol}(B_n(N))}{N^{\frac{n}{2}-1}} - \frac{S_\Gamma(N + \frac{1}{2}) - \mathrm{vol}(B_n(N + \frac{1}{2}))}{(N + \frac{1}{2})^{\frac{n}{2}-1}} \cdot (1 + \frac{1}{2N})^{\frac{n}{2}-1}.$$

This contradicts $\mathrm{vol}(B_n(N + \frac{1}{2})) - \mathrm{vol}(B_n(N)) = \Omega(N^{\frac{n}{2}-1})$.

$\square$

Although we can prove that a lattice with maximal complete lengths does not exist for any dimension $n \geq 2$, let us first consider the case $2 \leq n \leq 8$ to point out the essential difference between the knowledge in the first eight and in higher dimensions.

Then in the following section we discuss the next natural question, if there exist two lattices at all such that the complete lengths of one lattice dominate the complete lengths of the other.

3.2 Maximal complete lengths

3.2.1 Dimensions 2 to 8

Theorem 3.2.1. *In dimensions 2 to 8 a lattice with "maximal complete lengths" does not exist, i.e. there is no lattice such that for all $k \geq 0$ its k-th complete length is greater or equal than the k-th complete length of any other lattice in the same dimension with the same covolume.*

Proof. Since a lattice with maximal complete length spectrum must have a maximal first length the only possible lattices are the lattices with the best known sphere packings of Table 2.1. We know their first length and especially the multiplicity of this length. Therefore we can easily give a lattice that is not dominated by Σ_n. Although we could try any lattice (and probably we would succeed sooner or later) we will give a more general example:

Therefore let σ_1 be the first length of the normalised lattice Σ_n with the best known sphere packings in dimension $n \in \{2, ..., 8\}$ and $r(\sigma_1^2)$ its multiplicity cf. Section 2.5.1.

Remark 3.2.2. Once again we touch an old problem here, closely related to the packing problem: finding the maximum of the so called *kissing numbers* in any dimension, i.e. the number of spheres of a packing that touch one sphere. Since for lattice packings the kissing number is the same for every sphere it is equal to the multiplicity of the first length. Among lattices the optimal kissing numbers are known in dimension 1 to 8 and 24. It is less surprising that the lattices with the best known sphere packings realize these numbers for $n \leq 8$. Therefore $r(\sigma_1^2)$ is optimal among all lattices.
The general kissing number problem is only known in dimensions 1, 2, 3, 8 and 24, but in no other dimension cf. [CoSl4, Chap.1.2].

Table 3.1 gives the first length and the kissing number of Σ_n cf. [NeSl].

Table 3.1: Kissing numbers of the best known sphere packings

n	2	3	4	5	6	7	8
σ_1^2	$\sqrt{\frac{4}{3}}$	$\sqrt[3]{2}$	$\sqrt[4]{4}$	$\sqrt[5]{8}$	$\sqrt[6]{\frac{64}{3}}$	$\sqrt[7]{64}$	2
	≈ 1.15	≈ 1.26	≈ 1.41	≈ 1.52	≈ 1.67	≈ 1.81	2
$r(\sigma_1^2)$	6	12	24	40	72	126	240
$\sqrt[n]{\left(\frac{r(\sigma_1^2)}{2}\right)^2}$	3	≈ 3.30	≈ 3.46	≈ 3.31	≈ 3.30	≈ 3.26	≈ 3.31

Hence we are able to choose ε with $0 < \varepsilon < \sqrt[n]{\left(\frac{r(\sigma_1^2)}{2}\right)^2} - \sigma_1^2$. Furthermore for $2 \leq n \leq 8$ we define the orthogonal quadratic form

$$Q_{\Gamma_{n,\varepsilon}}(x) := x^T \cdot \begin{pmatrix} a & 0 & \cdots & 0 \\ 0 & \sqrt[n-1]{a^{-1}} & \ddots & \vdots \\ \vdots & \ddots & \ddots & 0 \\ 0 & \cdots & 0 & \sqrt[n-1]{a^{-1}} \end{pmatrix} \cdot x, \text{ with } a := \frac{\sigma_1^2 + \varepsilon}{\left(\frac{r(\sigma_1^2)}{2}\right)^2}.$$

By choice of ε we have $\sqrt[n-1]{a^{-1}} > \sigma_1^2 + \varepsilon$. And since $Q_{\Gamma_{n,\varepsilon}}$ is given in a reduced form, the $r(\sigma_1^2)$-th complete length of Γ is $\frac{r(\sigma_1^2) \cdot \sqrt{a}}{2}$ and therefore strictly greater than $\sigma_{r(\sigma_1^2)}^{(c)}$:

$$(\gamma_{r(\sigma_1^2)}^{(c)})^2 = \left(\frac{r(\sigma_1^2)}{2}\right)^2 \frac{\sigma_1^2 + \varepsilon}{\left(\frac{r(\sigma_1^2)}{2}\right)^2} = \sigma_1^2 + \varepsilon > \sigma_1^2 = (\sigma_{r(\sigma_1^2)}^{(c)})^2.$$

This proves Theorem 3.2.1. $\square$

3.2.2   General situation

Now we want to show the analogous result in any arbitrary dimension $n \geq 2$ but in contrast to Section 3.2.1 we do not know the biggest possible first length or its multiplicity. Nevertheless we are able to give the stronger result that even locally, i.e. in any small region around a given lattice, a lattice with maximal complete lengths does not exist.

For this purpose we use the well known fact that two different norms on $\mathbb{R}^n$ are equivalent:

Lemma 3.2.3. *Let Q be a positive definite form in $\mathbb{R}^n$ and denote the ordinary Euclidean length by $|\cdot|$. Then $Q(x) \geq c_Q |x|^2$ for all real vectors x, where $c_Q > 0$ is the lower bound of $Q(x)$ with $|x| = 1$.*

Proof. $Q(x)$ is continuous and so attains the lower bound c_Q on the compact set $|x| = 1$. Since Q is positive definite $c_Q > 0$. $\square$

Theorem 3.2.4. *Let Γ be a lattice in $\mathbb{R}^n$ with $n \geq 2$. Then there exists a lattice Γ' arbitrarily close to Γ such that the complete lengths of Γ do not dominate the complete lengths of Γ'.*

Proof. Let Γ be a lattice with a reduced basis $B = \{b_1, b_2, \ldots, b_n\}$. For $0 < \varepsilon < 1$ we define the lattice $\Gamma' := \Gamma_\varepsilon$ by its basis $B_\varepsilon := \{\varepsilon b_1, b_2, \ldots, b_{n-1}, \frac{1}{\varepsilon} b_n\}$. Obviously Γ_ε has the same covolume as Γ and $|\varepsilon b_1| < |b_2| \leq \ldots \leq |b_{n-1}| < |\frac{1}{\varepsilon} b_n|$, while Γ_ε is not necessarily reduced.

As usual a_{ij} are the entries of the matrix A_{Γ_B} of the associated form. We have for all $x \in \mathbf{Z}^n$:

$$d_\varepsilon(x) := Q_\Gamma(x) - Q_{\Gamma_\varepsilon}(x)$$

$$= (1 - \varepsilon^2)a_{11}x_1^2 + (1 - \frac{1}{\varepsilon^2})a_{nn}x_n^2 + 2(1 - \varepsilon)\sum_{j=2}^{n-1}a_{1j}x_1x_j + 2(1 - \frac{1}{\varepsilon})\sum_{j=2}^{n-1}a_{nj}x_nx_j.$$

For all real x with $|x| = 1$ we have $|x_j| \le 1$ for $j = 1, \dots, n$ and so

$$|d_\varepsilon(x)| \le (1 - \varepsilon^2)a_{11} + (1 - \frac{1}{\varepsilon^2})a_{nn} + 2(1 - \varepsilon)\sum_{j=2}^{n-1}|a_{1j}| + 2(1 - \frac{1}{\varepsilon})\sum_{j=2}^{n-1}|a_{nj}|.$$

Therefore we can choose $\varepsilon \in I := (b_\delta, 1)$ with $b_\delta \ne 1$ such that for all x with $|x| = 1$:

$$|d_\varepsilon(x)| \le \delta < c_{Q_\Gamma} \text{ with fixed } \delta \ne 0.$$

If $c_{Q_\Gamma} \ge c_{Q_{\Gamma_\varepsilon}}$ due to Lemma 3.2.3 there exists a $|x_0| = 1$ with $d_\varepsilon(x_0) = c_{Q_\Gamma} - Q_{\Gamma_\varepsilon}(x_0) \ge c_{Q_\Gamma} - c_{Q_{\Gamma_\varepsilon}} > 0$. Analogously if $c_{Q_\Gamma} < c_{Q_{\Gamma_\varepsilon}}$ we have $d_\varepsilon(x_0') \ge c_{Q_{\Gamma_\varepsilon}} - c_{Q_\Gamma} > 0$.

Hence: $\left|c_{Q_{\Gamma_\varepsilon}} - c_{Q_\Gamma}\right| \le \delta$ for all $\varepsilon \in I \Rightarrow c_{Q_{\Gamma_\varepsilon}} \ge c_{Q_\Gamma} - \delta > 0$ for all $\varepsilon \in I$. From Lemma 3.2.3 we know that for all x

$$Q_{\Gamma_\varepsilon}(x) \ge c_{Q_{\Gamma_\varepsilon}}|x|^2 \ge (c_{Q_\Gamma} - \delta)|x|^2. \quad (*)$$

Now assume that the complete lengths of Γ dominate the complete lengths of Γ_ε for all $\varepsilon \in I$. According by

$$\#(\Gamma \cap B_n(X)) \le \#(\Gamma_\varepsilon \cap B_n(X))$$

for all X. But by choice of Γ_ε we have $d_\varepsilon(e_n) < 0$, where e_n denotes the n-th standard basis vector. Hence there exists an $x_0 \in \mathbf{Z}^n$ such that on the one hand

$$Q_{\Gamma_\varepsilon}(x_0) \le Q_\Gamma(e_n) = |b_n|^2 \quad (**)$$

and on the other hand

$$|b_n|^2 < Q_\Gamma(x_0). \quad (***)$$

$\Rightarrow$ For all $\varepsilon \in I$ there exists an $x_0 \in \mathbf{Z}^n$ such that

$$\underset{(*)}{(c_{Q_\Gamma} - \delta)|x_0|^2} \le \underset{(**)}{Q_{\Gamma_\varepsilon}(x_0)} \le \underset{(***)}{|b_n|^2} < \min\left\{Q_\Gamma(x) > |b_n|^2\right\} \le Q_\Gamma(x_0).$$

$\Rightarrow$ For all $\varepsilon \in I$ there exists an $x_0 \in \mathbf{Z}^n$ such that

$$|x_0|^2 \le \frac{1}{c_{Q_\Gamma} - \delta}|b_n|^2 \text{ and } |d_\varepsilon(x_0)| \ge \min\left\{Q_\Gamma(x) > |b_n|^2\right\} - |b_n|^2 = \text{const} > 0.$$

Since $\frac{1}{c_{Q_\Gamma} - \delta}|b_n|^2$ does not depend on the choice of $\varepsilon \in I$ the set of all possible x_0 is finite. This is in contradiction to $\lim_{\varepsilon \to 1} d_\varepsilon(x) = 0$ for fixed x. $\qquad \square$

Remark 3.2.5. Since in the proofs of Theorem 3.2.1 and 3.2.4 a rational ε can be choosen as well, even among the set of arithmetic lattices the existence of a lattice with maximal complete lengths is excluded.

3.3   Further questions

3.3.1   Dominating complete lengths

In this section we will discuss the question if any lattice can dominate another arbitrary lattice, particularly we can ask:

Question 3.3.1. *If Γ and Γ' are two lattices with the same covolume such that the complete lengths of Γ dominate the complete lengths of Γ', do then all lengths of Γ and Γ' have to be equal?*

Remark 3.3.2. It is long known that 2-dimensional lattices with the same length spectrum are congruent. In higher dimensions this fact holds no longer true. The first pair of non-congruent lattices with the same (complete) lengths in dimension 4 was found by Schiemann [Schi1] in 1990. These examples have been generalized in [CoSl3]. Therefore we can easily give examples in every higher dimension simply by direct sums of lattices. In dimension 3 the problem was open till 1993. In his thesis [Schi2] Schiemann showed that a ternary integral lattice is determined by its theta function cf. Definition 3.3.4. Here it would be nice to find another proof that does not rely on a computer. Nevertheless, this means for Question 3.3.1 that in dimensions 1 to 3 the two lattices Γ and Γ' would be congruent.

In contrast to $P_\Gamma(X)$ the number $S_\Gamma(X)$ grows "uniformly". Hence if in dimension n we could answer Question 3.3.1 in the affirmative, it holds also true in every other dimension less than n:

Proposition 3.3.3. *If there exist n-dimensional lattices Γ and Γ' with the same covolume such that $\gamma_k^{(c)} \geq \gamma_k'^{(c)}$ for all $k \geq 0$ and $\gamma_{k_0}^{(c)} > \gamma_{k_0}'^{(c)}$ for at least one $k_0 \in \mathbb{N}$, then there exist lattices with the same properties in every higher dimension.*

Proof. Let Γ_n and Γ'_n be such n-dimensional lattices. There exists a bijective map $\varphi_n : \mathbb{Z}^n \to \mathbb{Z}^n$ such that $Q_{\Gamma_n}(x) \geq Q_{\Gamma'_n}(\varphi_n(x))$ for all $x \in \mathbb{Z}^n$ and $Q_{\Gamma_n}(x_0) > Q_{\Gamma'_n}(\varphi_n(x_0))$ for one x_0.
Define the $(n+1)$-dimensional lattices $\Gamma_{n+1} := \Gamma_n \perp \mathbb{Z}$ resp. $\Gamma'_{n+1} := \Gamma'_n \perp \mathbb{Z}$ and the bijective map $\varphi_{n+1} : \mathbb{Z}^{n+1} \to \mathbb{Z}^{n+1}$, $x \mapsto (\varphi_n(x_1, \ldots, x_n), x_{n+1})$.
Since $Q_{\Gamma_{n+1}}(x) = Q_{\Gamma_n}(x_1, \ldots, x_n) + x_{n+1}^2$ resp. $Q_{\Gamma'_{n+1}}(x) = Q_{\Gamma'_n}(x_1, \ldots, x_n) + x_{n+1}^2$ we have $Q_{\Gamma_{n+1}}(x) \geq Q_{\Gamma'_{n+1}}(\varphi_{n+1}(x))$ for all $x \in \mathbb{Z}^{n+1}$ and $Q_{\Gamma_{n+1}}(x_0, 0) > Q_{\Gamma'_{n+1}}(\varphi_{n+1}(x_0, 0))$. Hence the complete lengths of Γ_{n+1} are dominant and the length spectra are not equal. $\qquad\square$

3.3.2   Theta functions and modular forms

A useful approach to understand a sequence of numbers is to think of them as coefficients of some series. For our purpose we are interested in the *multiplicities* $r_\Gamma(m)$ of the appearing lengths in a lattice, i.e. the number of elements $x \in \mathbb{Z}^n$ such that $Q_\Gamma(x) = m$, cf. Section 2.5.1. Since in this context non-arithmetic lattices have not really a meaningful series we consider in this section only arithmetic respectively integral lattices. Due to Theorem 3.1.1 $r_\Gamma(m)$ is bounded by a polynomial in m. Thus one can define a function for every lattice on the upper half plane $\mathbb{H}$:

Definition 3.3.4. The function

$$\Theta_\Gamma(z) := \sum_{m=0}^\infty r_\Gamma(m) q^m, \quad \text{where } q = e^{2\pi i z}$$

is called the *theta function* of Γ. It converges for $|q| < 1$, hence it is holomorphic on $\mathbb{H}$.

In particular theta functions are examples of "modular forms".

Let the full modular group $\mathrm{SL}_2(\mathbb{Z})$ act on $\mathbb{H}$ as usual

$$A = \begin{pmatrix} a & b \\ c & d \end{pmatrix} \longmapsto \frac{az+b}{cz+d} =: A(z).$$

For a meromorphic function f this renders the operator $|_k$ for all $k \in \mathbb{Z}$

$$f|_k A(z) := (cz+d)^{-k} f\left(\frac{az+b}{cz+d}\right).$$

Furthermore denote the following congruence subgroup of $\mathrm{SL}_2(\mathbb{Z})$ for $N \in \mathbb{N}$

$$\Gamma_0(N) := \left\{ A = \begin{pmatrix} a & b \\ c & d \end{pmatrix} \in \mathrm{SL}_2(\mathbb{Z}) : c \equiv 0 \,(\mathrm{mod}\ N) \right\}.$$

Definition 3.3.5. A holomorphic function $f : \mathbb{H} \longrightarrow \mathbb{C}$ is called a *modular form* of weight $k \in \frac{1}{2}\mathbb{Z}$, level N for some character χ on $\mathbb{Z}/N\mathbb{Z}$ if

$$f|_k A(z) = \chi(d) f(z) \text{ for all } A \in \Gamma_0(N)$$

and $f|_k M(z)$ is bounded for $\mathrm{Im}(z) \to \infty$ for all $M \in \mathrm{SL}_2(\mathbb{Z})$.
If furthermore f vanishes at all cusps, i.e. $\lim_{t \to \infty} f|_k M(it) = 0$ for all $M \in \mathrm{SL}_2(\mathbb{Z})$, it is called a *cusp form*.

Since $\begin{pmatrix} 1 & 1 \\ 0 & 1 \end{pmatrix} \in \Gamma_0(N)$, we see that $f(z) = f(z+1)$. A modular form is thus given by a Fourier expansion

$$f(z) = \sum_{m=0}^\infty a(m) q^m = \sum_{m=0}^\infty a(m) e^{2\pi i m z}$$

with Fourier coefficients $a(m) \in \mathbb{C}$.

Now we have to put it more precisely, cf. [An, Thm.2.2.2]:

Theorem 3.3.6. *Let Γ be an even n-dimensional lattice. Then Θ_Γ is a modular form (with integral Fourier coefficients) of weight $\frac{n}{2}$, level N_Γ, where N_Γ is the least positive integer such that the matrix $N_\Gamma A_\Gamma^{-1}$ is even. And the character χ_Γ only depends on n and $\det(Q_\Gamma)$.*

Hence for integral lattices Γ and Γ' with the same covolume

$$\Theta_{\Gamma'\Gamma}(z) = \sum_{m=0}^{\infty} r_{\Gamma'\Gamma}(m)q^m := \Theta_{\Gamma'}(z) - \Theta_{\Gamma}(z)$$

is a modular form of level $N_{\Gamma'\Gamma}$ (the least common multiple of $N_{\Gamma'}$ and N_Γ) with integral Fourier coefficients. If now the lengths of Γ dominate we have for all X in $\mathbb{N}$

$$\sum_{m=0}^{X} r_{\Gamma'}(m) \geq \sum_{m=0}^{X} r_{\Gamma}(m),$$

and respectively

$$\sum_{m=0}^{X} r_{\Gamma'\Gamma}(m) \geq 0.$$

Thus for integral lattices we can also state Question 3.3.1 in more general in terms of modular forms:

Question 3.3.7. *Do there exist two different modular forms*
$f(z) = \sum_{m=0}^{\infty} a(m)q^m$ *and* $f'(z) = \sum_{m=0}^{\infty} a'(m)q^m$ *(with integral coefficients) of the same weight and level such that for all $X \in \mathbb{N}$*

$$\sum_{m=0}^{X} (a'(m) - a(m)) \geq 0 ?$$

It seems that this is a difficult question in general, but we are able to answer in the negative in the (very) special case of even unimodular lattices.

3.3.3   Even unimodular lattices

Obviously the level of the theta function Θ_Γ of an even unimodular lattice Γ is 1. Hence Θ_Γ is a modular form for the full $\mathrm{SL}_2(\mathbb{Z})$ with $r_\Gamma(0) = 1$ (and so $r_{\Gamma'\Gamma}(0) = 0$). Furthermore it is not difficult to see that the dimension of Γ must be divisible by 8 [Se1, p.109]. Therefore $\Theta_{\Gamma'\Gamma}$ is a cusp form for $\mathrm{SL}_2(\mathbb{Z})$ of integral weight. The coefficients of integral weight cusp forms are well studied:

Estimates for the coefficients of cusp forms
Let f be a cusp form of integral weight k. We are interested in the growth of the coefficients $a(m)$. A first bound is

$$a(m) = \mathcal{O}(m^{\frac{k}{2}}).$$

This bound is often referred to as the trivial bound for cusp forms. The exponent $\frac{k}{2}$ can be improved. Deligne has shown in [De, p.302] that

$$a(m) = \mathcal{O}(m^{\frac{k}{2}-\frac{1}{2}} \cdot \#\{d \in \mathbb{N} : d|m\}).$$

This clearly implies that

$$a(m) = \mathcal{O}(m^{\frac{k}{2}-\frac{1}{2}+\varepsilon}) \quad \text{for every } \varepsilon > 0.$$

To achieve the end that we have in view estimates for sums of coefficients of cusp forms are useful. Due to Deligne [De] we know that

$$\sum_{m=0}^{X} a(m) = \mathcal{O}(X^{\frac{k}{2}-\frac{1}{6}+\varepsilon}) \quad \text{for every } \varepsilon > 0.$$

In the other direction Walfisz showed [Wa, p.76]

$$\sum_{m=0}^{X} a(m) = \Omega(X^{\frac{k}{2}-\frac{1}{4}}).$$

This estimate has been improved by a number of authors, see [Ran] and the references specified therein. For example Hafner and Ivić [HaIv, Thm.3] have shown that, for some positive constant D,

$$\sum_{m=0}^{X} a(m) = \Omega_{\pm}\left(X^{\frac{k}{2}-\frac{1}{4}} \cdot \exp\left(D \cdot \frac{(\log\log(X))^{\frac{1}{4}}}{(\log\log\log(X))^{\frac{3}{4}}} \right) \right).$$

Hence for $\varepsilon > 0$ there exist positive c_1 and c_2 such that for infinitely many $X, X' \in \mathbb{N}$

$$\sum_{m=0}^{X} a(m) \geq c_1 X^{\frac{k}{2}-\frac{1}{4}-\varepsilon} \quad \text{and} \quad \sum_{m=0}^{X'} a(m) \leq -c_2 X'^{\frac{k}{2}-\frac{1}{4}-\varepsilon}.$$

Additionally since $a(X)$ is at most of order $X^{\frac{k}{2}-\frac{1}{2}+\varepsilon}$ we know that there are infinitely many partial sums between these two orders. However, in particular it follows that the partial sums change sign infinitely often, and that is all we need to answer Question 3.3.1 for even unimodular lattices.

Corollary 3.3.8. *Let Γ and Γ' be two even unimodular lattices in the same dimension with different theta functions. Then infinitely many complete lengths of Γ are greater and infinitely many complete lengths are less than the complete lengths of Γ'.*

Hence we see that even an asymptotical dominance is excluded. This motivates the guess that Question 3.3.1 has to be answered in the affirmative for all lattices, but mind, unimodular and even are serious restrictions. However, there are some (nice) examples.

Examples

In dimension 8 the only even unimodular lattice is E_8, so there is nothing to examine here. In dimension 16 we have the two non-equivalent lattices $E_8 \times E_8$ and D_{16}, but they have the same theta function.

The first interesting examples are in dimension 24. There exist 24 even unimodular lattices, listed by Niemeier [Ni], 23 with minimal norm 2, and one, the Leech lattice Λ_{24}, with minimal norm 4. The space of modular forms of weight 12 is of dimension 2. Thus the theta function of an even unimodular 24-dimensional lattice Γ can be written by a combination of 2 basis functions. For the coefficients $r_\Gamma(2m)$ of Θ_Γ we have:

$$r_\Gamma(2m) = \frac{65520}{691} \cdot \sigma_{11}(m) + c_\Gamma \cdot \tau(m),$$

where $\sigma_j(m) := \sum_{d|m} d^j$ is the sum of the j-th powers of positive divisors of m and $\tau(m)$ is the m-th coefficient of the cusp form given by

$$T(z) = q \cdot \prod_{m=1}^{\infty} (1 - q^m)^{24} =: \sum_{m=1}^{\infty} \tau(m) \cdot q^m.$$

The function $m \mapsto \tau(m)$ is called the *Ramanujan τ-function*. The lattice constant c_Γ is determined by

$$c_\Gamma = r_\Gamma(1) - \frac{65520}{691},$$

for more information see [Se1, Chap.6.6.]. Hence the multiplicities $r_\Gamma(m)$ and $r_{\Gamma'}(m)$ of two such lattices Γ and Γ' just differ in a factor of the τ-function. Since $T(z)$ is a cusp form, the *summatory τ-function*

$$\Sigma\tau : X \mapsto \sum_{m=1}^{X} \tau(m)$$

changes sign infinitely often. Obviously these changes of sign indicate where the dominance of lengths is changing:

$$\sum_{m=1}^{X} r_{\Gamma'\Gamma}(2m) \geq 0 \;\Leftrightarrow\; \sum_{m=1}^{X} (c_{\Gamma'} - c_\Gamma) \cdot \tau(m) \geq 0.$$

Assume $c_{\Gamma'} > c_\Gamma$, if now $\Sigma\tau(X-1) < 0$ and $\Sigma\tau(X) > 0$ then there exists a number k such that $\sqrt{2X} = \gamma_k^{(c)} > \gamma_k'^{(c)}$. And further if there is no change of sign at X and $\Sigma\tau(X) > 0$, then $\gamma_k^{(c)} \geq \gamma_k'^{(c)}$ for all k with $\gamma_k^{(c)} = \sqrt{2X}$.

The first values of $\Sigma\tau$ are:

1, -23, 229, -1243, 3587, -2461, -19205, 65275, -48368, -164288, 370324, -620, -578358, -176502, 1040658, 2027794, -4878140, -2150708, 8510712, 1400952, -2818536, -15649224, 2994048, 24283008,

$-1216217, 12649495, -60629585, -35982417, 92424213, 63212373, 10369205, -186337099, -51614875,$
$114127541, 33254021, 200536517, 18323203, -237550877, -383140853, 24897547, 333017989, ...$

Hence the first changes of sign of $\Sigma\tau$ are at:
X= 2, 3, 4, 5, 6, 8, 9, 11, 12, 15, 17, 19, 21, 23, 25, 26, 27, 29, 32, 34, 38, 40, 44, 47, 49, 50, 51, 55,
59, 61, 67, 70, 76, 79, 83, 88, 93, 97, 99, 100, 103, 108, 113, 119, 125, 131, 136, 141, 144, 145, 149, 153,
160, 167, 173, 179, 187, 193, 201, 206, 208, 209, 211, 212, 216, 223, 229, 232, 233, 236, 245, 251, 252,
253, 256, 257, 262, 269, 277, 286, 287, 288, 295, 303, 313, 319, 324, 325, 331, 333, 334, 338, ...

Since the Leech lattice Λ_{24} is the only one of the 24 lattices mentioned above, which
does not represent 2, it has the smallest factor $c_\Lambda = -\frac{65520}{691}$.
Hence we know certainly that its k-th length $\lambda_k^{(c)}$ is greater than or equal to the k-th
length of all other even unimodular 24-dimensional lattices if

$$(\lambda_k^{(c)})^2 = 32, 40, 48, 60, 62, 70, 72, 74, 82, 84, 86, 96, 112, 114, 116, 124, ...$$

and less or equal if

$$(\lambda_k^{(c)})^2 = 14, 20, 26, 28, 36, 44, 56, 66, 78, 90, 92, 104, 106, 108, 120, 136, ...$$

This holds similarly for other lattices, for example for $E_8 \times E_8 \times E_8$ and D_{24}, since
$c_{E_8^3} = \frac{432000}{691} < \frac{697344}{691} = c_{D_{24}}$.

Index

B_n, 11, 39

complete length, 39
covolume cov(Γ), 5
cusp form, 45

$\delta(Q)$, 25

Erdös number E_Γ, 15

Gamma function Γ, 10, 40
genus gen(Q), 7

kissing number, 41

lattice, 5
    A_n, D_n, E_6, E_7, E_8, 11
    face-centered cubic Σ, 14, 16
    honeycomb Λ, 17
    maximal lengths, 13
    reduced, 6
length, 13
local-global principle, 7

modular form, 45
multiplicity $r(\gamma_k^2)$, 25, 44

population function P_Γ, 15

quadratic form
    (classically) integral, 5
    equivalent, 5
    arithmetic, 6
    even, 6
    irregular, 21
    locally equivalent, 7
    Ramanujan's form, 22
    reduced, 6
    universal, 8

Ramanujan τ-function, 48

S_Γ, 39
Schmutz Schaller conjecture, 13
sphere packing, 11, 13

theta function Θ_Γ , 45

Bibliography

[An]   A. N. Andrianov. *Quadratic Forms and Hecke Operators.* Grundlehren der Mathematischen Wissenschaften, **286**, Springer (1987).

[Bh]   M. Bhargava. *On the Conway-Schneeberger Fifteen Theorem.* Quadratic forms and their applications (Dublin, 1999), Contemp. Math., **272**, Amer. Math. Soc., Providence, RI (2000), 27-37.

[Ca]   J. W. S. Cassels. *Rational Quadratic Forms.* Academic press (1978).

[CoSl1] J. H. Conway, N. J. A. Sloane. *Low-dimensional lattices. IV. The mass formula.* Proc. Roy. Soc. London Ser. A, **1857** (1988), 259-286.

[CoSl2] J. H. Conway, N. J. A. Sloane. *Lattices with Few Distances.* J. Number Theory, **39** (1991), 75-90.

[CoSl3] J. H. Conway, N. J. A. Sloane. *Four-Dimensional Lattices with the same Theta Series.* Internat. Math. Res. Notices, **4** (1992), 93-96.

[CoSl4] J. H. Conway, N. J. A. Sloane. *Sphere Packings, Lattices and Groups.* Grundlehren der Mathematischen Wissenschaften, **290**, Springer 3. ed. (1999).

[De]   P. Deligne. *La conjecture de Weil I.* Inst. Hautes Etudes Sci. Publ. Math., **43** (1974), 273-307.

[Di1]   L. E. Dickson. *Integers Represented by Positive Ternary Quadratic Forms.* Bull. Amer. Math. Soc., **33** (1927), 63-70.

[Di2]   L. E. Dickson. *Quaternary Quadratic Forms Representing all Integers.* Amer. J. Math., **49** (1927), 35-56.

[DuSP] W. Duke, R. Schulze-Pillot. *Representation of integers by positive ternary quadratic forms and equidistribution of lattice points on ellipsoids.* Invent. Math., **99** (1990), 49-57.

[Ei]   G. Eisenstein. *Tabelle der reducirten positiven ternären quadratischen Formen, nebst den Resultaten neuer Forschungen über diese Formen, in besonderer Rücksicht auf ihre tabellarische Berechnung.* Journal für Mathematik, **41** (1851), 141-190.

[Fr]   F. Fricker. *Einführung in die Gitterpunktlehre*. LMW Mathematische Reihe **73**, Birkhäuser (1983).

[Ga]   C. F. Gauß. *Besprechung des Buchs von L. A. Seeber: Untersuchungen über die Eigenschaften der positiven ternären quadratischen Formen usw.* Göttingsche Gelehrte Anzeigen (1831) bzw. Werke II (1876), 188-196.

[HaIv]   J. L. Hafner, A. Ivić. *On sums of Fourier coefficients of cusp forms*. Enseign. Math.(2), **35** (1989), 375-382.

[Ha]   J. Hanke. *Some Recent Results about (Ternary) Quadratic Forms*. CRM Proc. Lecture Notes, **36**, Amer. Math. Soc., Providence, RI (2004), 147-164.

[Jo]   B. W. Jones. *The Arithmetic Theory of Quadratic Forms*. Carus Math. monographs, **10** (1961).

[Ka]   I. Kaplansky. *The First Nontrivial Genus of Positive Definite Ternary Forms*. Math. Comp., **64** (1995), 341-345.

[Kn]   M. Kneser. *Quadratische Formen*. Springer (1999).

[Kü1]   S. Kühnlein. *Partial Solution of a Conjecture of Schmutz*. Arch. Math., **67** (1996), 164-172.

[Kü2]   S. Kühnlein. *Multiplicities of Non-arithmetic Ternary Quadratic Forms and Elliptic Curves of Positive Rank*. Math. Z., **230** (1999), 529-543.

[MoRi]   P. Moree, H. J. J. te Riele. *The Hexagonal versus the Square Lattice*. Math. Comp., **73** (2004), 451-473.

[MoOs]   P. Moree, R. Osburn. *Two-dimensional Lattices with Few Distances*. Enseign. Math., **52** (2006), 361-380.

[NeSl]   G. Nebe, N. J. A. Sloane. *A Catalogue of Lattices*. data-base, http://www.research.att.com/~njas/lattices.

[NeXi]   G. Nebe, C. Xing. *A Gilbert-Varshamov type bound for Euclidean packings*. Math. Comp., **77** (2008), no. 264, 2339-2344.

[Ni]   H.-V. Niemeier. *Definite quadratische Formen der Dimension 24 und Diskriminante 1*. JNT, **5** (1973), 142-178.

[OnSo]   K. Ono, K. Soundararajan. *Ramanujan's ternary quadratic form*. Invent. Math., **130** (1997), 415-454.

[Ram]   S. Ramanujan. *On the Expression of a Number in the Form $ax^2+by^2+cz^2+dw^2$*. Math. Proc. Cambridge Philos. Soc., **19** (1917), 11-21.

[Ran]   R. A. Rankin. *Sums of cusp form coefficients*. Automorphic forms and analytic theory (Montreal, PQ 1989) Univ. Montréal, (1990), 115-121.

[Schi1] A. Schiemann. *Ein Beispiel positiv definiter quadratischer Formen der Dimension 4 mit gleichen Darstellungszahlen.* Arch. Math., **54** (1990), 372-375.

[Schi2] A. Schiemann. *Ternäre positiv definite quadratische Formen mit gleichen Darstellungszahlen.* Dissertation, Universität Bonn 1993. Bonner Mathematische Schriften, **268** (1994).

[Schm1] P. Schmutz. *Arithmetic Groups and the Length Spectrum of Riemann Surfaces.* Duke Math. J., **84** (1996), 199-215.

[Schm2] P. Schmutz Schaller. *Geometry of Riemann Surfaces Based on Closed Geodesics.* Bull. Amer. Math. Soc., **35** (1998), 193-214.

[Se1] J.-P. Serre. *A Course in Arithmetic.* Graduate Texts in Mathematics, **7**, Springer (1973).

[Se2] J.-P. Serre. *Complex semisimple Lie algebras.* Springer (1987).

[Si] C. L. Siegel. *Über die analytische Theorie der quadratischen Formen.* Ann. of Math., **36** (1935), 527-606.

[Wa] A. Walfisz. *Über die Koeffizientensummen einiger Modulformen.* Math. Ann., **108** (1933), 75-90.

[Wi] T. A. Willging. *On a conjecture of Schmutz.* Arch. Math., **91** (2008), 323-329.